T0413180

Ecological Studies, Vol. 214

Analysis and Synthesis

Ecological Studies

Further volumes can be found at springer.com

Volume 196
Western North American Juniperus Communites: A Dynamic Vegetation Type (2008)
O. Van Auken (Ed.)

Volume 197
Ecology of Baltic Coastal Waters (2008)
U. Schiewer (Ed.)

Volume 198
Gradients in a Tropical Mountain Ecosystem of Ecuador (2008)
E. Beck, J. Bendix, I. Kottke,
F. Makeschin, R. Mosandl (Eds.)

Volume 199
Hydrological and Biological Responses to Forest Practices: The Alsea Watershed Study (2008)
J.D. Stednick (Ed.)

Volume 200
Arid Dune Ecosystems: The NizzanaSands in the Negev Desert (2008)
S.-W. Breckle, A. Yair,
and M. Veste (Eds.)

Volume 201
The Everglades Experiments: Lessons for Ecosystem Restoration (2008)
C. Richardson (Ed.)

Volume 202
Ecosystem Organization of a Complex Landscape: Long-Term Research in the Bornhöved Lake District, Germany (2008)
O. Fränzle, L. Kappen, H.-P. Blume,
and K. Dierssen (Eds.)

Volume 203
The Continental-Scale Greenhouse Gas Balance of Europe (2008)
H. Dolman, R.Valentini, and A. Freibauer (Eds.)

Volume 204
Biological Invasions in Marine Ecosystems: Ecological, Management, and Geographic Perspectives (2009)
G. Rilov and J.A. Crooks (Eds.)

Volume 205
Coral Bleaching: Patterns, Processes, Causes and Consequences
M.J.H van Oppen and J.M. Lough (Eds.)

Volume 206
Marine Hard Bottom Communities: Patterns, Dynamics, Diversity, and Change (2009)
M. Wahl (Ed.)

Volume 207
Old-Growth Forests: Function, Fate and Value (2009)
C. Wirth, G. Gleixner,
and M. Heimann (Eds.)

Volume 208
Functioning and Management of European Beech Ecosystems (2009)
R. Brumme and P.K. Khanna (Eds.)

Volume 209
Permafrost Ecosystems: Siberian Larch Forests (2010)
A. Osawa, O.A. Zyryanova, Y. Matsuura,
T. Kajimoto, R.W. Wein (Eds.)

Volume 210
Amazonian Floodplain Forests: Ecophysiology, Biodiversity and Sustainable Management (2010)
W.J. Junk, M.T.F. Piedade, F. Wittmann,
J. Schöngart, P. Parolin (Eds.)

Volume 211
Mangrove Dynamics and Management in North Brazil (2010)
U. Saint-Paul and H. Schneider (Eds.)

Volume 212
Forest Management and the Water Cycle: An Ecosystem-Based Approach (2011)
M. Bredemeier, S. Cohen, D.L. Godbold,
E. Lode, V. Pichler, P. Schleppi (Eds.)

Volume 213
The Landscape Ecology of Fire (2011)
D. McKenzie, C.S. Miller, D.A. Donald (Eds.)

Volume 214
Human Population: Its Influences on Biological Diversity (2011)
R.P. Cincotta and L.J. Gorenflo (Eds.)

R.P. Cincotta • L.J. Gorenflo
Editors

Human Population

Its Influences on Biological Diversity

Editors
Richard P. Cincotta
Demographer-in-residence
The Stimson Center
1111 19th St., NW
12th Floor
Washington, DC 20036, USA
rcincotta@stimson.org

Larry J. Gorenflo
Department of Landscape Architecture
Stuckeman School of Architecture and
Landscape Architecture
121 Stuckeman Family Building
The Pennsylvania State University
University Park, PA 16802-1912, USA
ljg11@psu.edu

ISSN 0070-8356
ISBN 978-3-642-16706-5 e-ISBN 978-3-642-16707-2
DOI 10.1007/978-3-642-16707-2
Springer Heidelberg Dordrecht London New York

Library of Congress Control Number: 2011921262

Cover design: deblik Berlin, Germany

Printed on acid-free paper

Springer is part of Springer Science+Business Media (www.springer.com)

Dedication: A.J. Lotka

The career of Alfred J. Lotka (1880–1949) provides an instructive metaphor for the interdisciplinary demands of exploring the relationships between humans and the other species with which we share the thin "living surface" of this planet. Although Lotka, during the prime of his professional career, contributed to the computational foundations of several then-growing fields, his most influential accomplishments live on today in the overlapping theoretical and methodological lattices of two population fields: *demography*, the study of the characteristics and dynamics of human population; and *population biology*, the study of the population characteristics and dynamics of all other species. In population biology, Lotka wrote basic systems of nonlinear differential equations that characterize the dynamics of populations of predator and prey species and populations of competing species.

Given the long list of Lotka's intellectual contributions, it seems curious that he spent only 2 years in the employ of academe. From 1922 to 1924, Lotka held an appointment as a research fellow at Johns Hopkins University during which he completed his book *Elements of Mathematical Biology*. Having completed a manuscript setting down a series of mathematical approaches to evolutionary and population biology, Lotka found very little interest in his research in university departments of the day. Instead, he went to work for the Metropolitan Life Insurance Company, in New York, where he spent the final two and a half decades of his career improving the company's capacity for actuarial research. In this capacity, Lotka continued to identify and address applied problems and to publish research and hold key positions in professional societies that allowed him to influence the progressive and intertwining paths of demography, economics, and epidemiology.

Preface

The 1980s, the years during which we, the editors, conducted graduate research, represented the twilight of an era in which ecologists typically traveled far afield, to national parks and preserves, remote forests and far-off deserts, distant highlands, and isolated coral reefs, to identify the web of relationships that held together pristine ecological communities. Most investigators returned with news that their study community was, in fact, less pristine than hoped. Its initial appearance belied evidence, uncovered during intensive fieldwork, of anthropogenic disturbance – most often the intrusion of exotic species and diseases, and sometimes the more obvious impacts of human hunting and other types of exploitation, physical habitat alteration, the spread of manmade pollutants, or concentrations of naturally occurring compounds or native biota far in excess of background rates.

Two decades earlier, many cultural anthropologists still sought pristine human communities as their subjects of study. Just as ecologists had encountered overwhelming evidence of human presence in the biosphere, fieldwork brought anthropologists face to face with the intrusion of the state and its institutions, the proliferation of modern communications and other technologies, the influence of other *more developed* societies, and the powerful hand of the global economy. By the 1980s, many effects of broad acculturation had played out. Language, one of the indicators of cultural integrity, continued to reveal the impact of modern society as native tongues were replaced by widely spoken languages that increasingly dominated the world. Although the indicators of impacts were different, pristine cultures had become as rare, or rarer, than pristine ecosystems as the influence of the modern world expanded.

Reactions by ecologists and anthropologists to the decline and disappearance of ecosystems and cultures that were uninfluenced, directly or indirectly, by the expanding footprint of modern industrial society varied. Some renewed their search for such unaffected systems by traveling even farther from the beaten path to increasingly remote locations in rain forests, mountains, deserts, and oceans. Others used their scholarship as an entrée into environmental and human rights activism, hoping to reverse a downward spiral among endangered species and cultures.

This book emerges out of a third scholarly reaction to the disappearance of nature. This "third way" is embodied in the assumption that *Homo sapiens*, because of its unusually large, often dense, and increasingly widespread population, and because of the nearly limitless reach of human institutions, technologies and trade, should be considered a component of nearly all ecological systems and studied as such. Many of us who take this position were informed by the research of Paul Martin, which lent substantial evidence to the hypothesis that our species had a powerful hand in the widespread megafaunal extinctions of the Pleistocene Epoch. Others acknowledge the influence of cultural ecology developed by Julian Steward and successors, and the argument that human cultural behavior in part reacts to, and in turn influences, surrounding ecosystems.

In the twenty-first century and beyond, researchers must accept that each biotic population – regardless of body size, habitat, or distribution – exists in an institutional–biotic–abiotic space which extends from the global economy to the state and its institutions, to local populations of humans and their community institutions, and finally to the abiotic and biotic environment. Pressures that induce behavioral adaptation, genetic selection, and population change (including extinction) could emanate from any part of this multidimensional space.

In our instructions to authors, we asked them to review, within their respective fields of expertise, established and incipient theories – including their own novel theories – that strive to explain some aspect of the interaction between our planet's supply of biological diversity and a geographic or demographic aspect of human population (e.g., size, density, growth, age structure, or distribution). Beyond their review, authors were asked to describe key methodologies, to identify gaps where theory and empirical information is lacking, and to identify new research questions, where they arise. The editors wish to thank the authors for their efforts and contributions, their responses to reviews and edits, and their patience through an overly lengthy process of assembling, modifying, and rewriting manuscripts.

In addition, we thank Andrea Schlitzberger, the editor of the Ecological Studies Series, for her guidance and patience and for her enthusiasm for including a volume on human population in this series. We are also grateful for the efforts of Martyn Caldwell, Springer's local editor in the United States, who gave the chapters a thorough reading and contributed timely edits, perceptive thoughts, and useful recommendations. Finally, we thank a long list of anonymous reviewers, each of whom evaluated a chapter and together helped to generate a much improved volume.

Richard Cincotta would like to thank The Stimson Center (Ellen Laipson, President) and Environmental Change and Security Project (Geoffrey Dabelko, Director) at the Woodrow Wilson International Center for Scholars (both in Washington, DC, USA) for their support during this project. I thank Preminda Jacob, my spouse, and my daughter, Malayika Cincotta – both of whom have their own (more exciting) careers to worry about – for allowing me to shirk some of my family responsibilities to help complete this volume.

Larry Gorenflo would like to express his thanks to Elizabeth Hocking and Argonne National Laboratory, Keith Alger and Conservation International, and

Kelleann Foster, Brian Orland, and the Department of Landscape Architecture at Penn State University for their support while assembling this volume. Grace, my wife, and my daughters, Maria, Annie, and Lauren, showed their characteristic mixture of love and understanding during this project, proving the time and support necessary to finish the volume while tolerating all-too-frequent bouts of physical and mental absence.

Washington, DC, USA
University Park, PA USA
Fall 2010

Richard Cincotta
L.J. Gorenflo

Contents

Contributors

Robin Abell Conservation Science Program, World Wildlife Fund, 1250 24th St. NW, Washington, DC 20037, USA, robin.abell@wwfus.org

Li An Department of Geography, San Diego State University, San Diego, CA 92182, USA, lan@mail.sdsu.edu

Richard E. Bilsborrow Department of Biostatistics, University of North Carolina at Chapel Hill, Chapel Hill, NC 27599, USA, Richard_Bilsborrow@unc.edu

Erica N. Chambers Department of Anthropology, The Ohio State University, Columbus, OH 43210, USA, chambers.166@osu.edu

Kenneth M. Chomitz Independent Evaluation Group, The World Bank, 1818 H Street, NW Washington, DC 20036, USA, kchomitz@worldbank.org

Richard P. Cincotta Demographer-in-residence, H.L. Stimson Center, 1111 19th Street, 12th Floor, NW, Washington, DC 20036, USA, rcincotta@stimson.org

Catherine Corson Program in Environmental Studies, Mount Holyoke College, 50 College Street, South Hadley, MA 01075, USA, ccorson@mtholyoke.edu

Thomas W. Crawford Department of Geography, Center for Geographic Information Science, East Carolina University, Greenville, NC 27858, USA, crawfordt@ecu.edu

Robert Engelman Worldwatch Institute, 1776 Massachusetts Ave, NW Washington, DC 20036-1904, USA, rengelman@worldwatch.org

Stuart R. Gaffin Center for Climate Systems Research, Columbia University, 2880 Broadway, New York, NY 10025, USA, stuart.gaffin@gmail.com

L.J. Gorenflo Department of Landscape Architecture, Pennsylvania State University, University Park, PA 16802, USA, lgorenflo@psu.edu

Lee Hachadoorian Earth & Environmental Sciences, The Graduate Center of the City University of New York, 365 Fifth Ave, New York, NY 10016, USA, lee.hachadoorian@gmail.com

Grady Harper 4165 Downy Court, East Helena, MT 59635, USA, gradyharper@yahoo.com

Guangming He Center for Systems Integration and Sustainability, Department of Fisheries and Wildlife, Michigan State University, East Lansing, MI 48824, USA, heguangm@msu.edu

Miroslav Honzák Science and Knowledge Division, Conservation International, 2011 Crystal Drive, Suite 500, VA 22202, USA, m.honzak@conservation.org

Bernhard Lehner Department of Geography, McGill University, 805 Sherbrooke Street, West Montreal, QC, Canada H3A 2K6, bernhard.lehner@mcgill.ca

Marc Linderman Center for Systems Integration and Sustainability, Department of Fisheries and Wildlife, Michigan State University, East Lansing, MI 48824, USA, marc-linderman@uiowa.edu

Jianguo Liu Center for Systems Integration and Sustainability, Department of Fisheries and Wildlife, Michigan State University, East Lansing, MI 48824, USA, jliu@panda.msu.edu

Flora Lu Department of Latin American and Latino Studies, University of California, Santa Cruz, CA 95064, USA, floralu@ucsc.edu

Deirdre Mageean Division of Research and Graduate Studies, East Carolina University, Greenville, NC 27858, USA, mageeand@ecu.edu

Jeffrey K. McKee Department of Anthropology, The Ohio State University, Columbus, OH 43210, USA, mckee.95@osu.edu

Zhiyun Ouyang Key Lab of Systems Ecology, Chinese Academy of Sciences, Beijing 100085, People's Republic of China, zyouyang@rcees.ac.cn

Berk Özler Development Research Group, The World Bank, 1818 H Street, NW, Washington, DC 20036, USA, bozler@worldbank.org

Steward Pickett Cary Institute of Ecosystem Studies, Box AB, 2801 Sharon Turnpike, Millbrook, NY 12545-0129, USA, PickettS@CaryInstitute.org

Christopher Small Lamont-Doherty Earth Observatory of Columbia University, Palisades, NY 10964-1000, USA, small@Ldeo.columbia.edu

Katalin Szlavecz Department of Earth & Planetary Sciences, Olin Hall, Johns Hopkins University, 3400 N. Charles Street, Baltimore, MD 21218-2687, USA, szlavecz@jhu.edu

Michele Thieme Conservation Science Program, World Wildlife Fund, 1250 24th St. NW, Washington, DC 20037, USA, michele.thieme@wwfus.org

Paige Warren Department of Environmental Conservation, Holdsworth Hall, University of Massachusetts – Amherst Holdsworth Way, Amherst, MA 01003, USA, pswarren@nrc.umass.edu

Chapter 1
Introduction: Influences of Human Population on Biological Diversity

Richard P. Cincotta and L.J. Gorenflo

1.1 Introduction

This volume is the result of an effort to document the state of research and promote progress in addressing a fundamental question facing conservationists: What can the dynamics of *Homo sapiens*' demographic and geographic distributions tell researchers and conservation practitioners about the status of, and prospects for, biological diversity? This question lies at the core of a field of study that, despite its logic and promise, remains remarkably underdeveloped. Yet finding answers to that question and applying general lessons learned from the answers may well be essential to maintain what remains of Earth's biological diversity.

From our vantage point, biodiversity conservation may have only modest hope of achieving any measurable degree of success. Human activity currently consumes roughly 40% of our planet's *annual gross terrestrial primary productivity* (the biomass produced through photosynthesis). Our species has already converted almost one-third of the terrestrial surface to agricultural fields and urban areas (United Nations Development Program et al. 2000; Vitousek et al. 1997). And this wholesale transformation of our planet's biosphere is anticipated to continue at an alarming rate. Global human population, which reached the 2.5 billion mark in 1950, has been estimated by the United Nations Population Division (2009) at about 6.9 billion in mid-2010. According to their most recent revision, UN demographers project that by 2050, human population will range between 7.8 billion (the UN low fertility variant) and 10.8 billion (high fertility variant), with a best guess of about 9.2 billion (the UN medium fertility variant). Unlike the past, much of this growth will occur in the humid tropics in and around ecosystems that support the planet's richest concentrations of endemic species (Cincotta and Engelman 2000; Cincotta et al. 2000). It is reasonable to expect that these people's rights to decent housing and adequate nutrition, clean water, sufficient energy, and a means to participate in their state's economy will supersede efforts to protect nonhuman species from extinction. Several researchers have reported relationships suggesting various influences of human population density, growth, and migration on biodiversity, at the global and regional scale (Gorenflo 2002; Holdaway and Jacomb 2000; Martin 1984; McKee et al. 2003), and the composition of biological communities, at the

R.P. Cincotta and L.J. Gorenflo (eds.), *Human Population: Its Influences on Biological Diversity*, Ecological Studies 214,
DOI 10.1007/978-3-642-16707-2_1, © Springer-Verlag Berlin Heidelberg 2011

local scale (Hoare and Du Toit 1999). The research in this volume provides further evidence on how the settlement patterns of humans, the densities at which they live, and the ways they connect these settlements to others set limits on the richness of species on Earth and pose challenges to attempts to conserve them.

1.2 Objectives

In assembling this volume, we endeavored to review a significant breadth of the human geographic and biogeographic research literature that addresses the relationships between human settlement, locally native species, and their ecosystems. This breadth emerges as analyses at three geographic scales: the global scale, the level of the biome, and the level of local ecosystems. Approaches range along a spectrum, from work that is almost purely theoretical to highly empirical research that organizes and analyzes available data.

In writing their chapters, the authors of this volume were asked to review relevant theories, both established and those arising from their own research, and to identify empirical indicators and observations that they found germane to investigating relationships between populations of native species and aspects of human demography and geography. We charged them to suggest where such theories and observations should lead researchers in the future and to identify gaps where theory and empirical reinforcement appeared to be lacking. Both social and biological scientists have contributed chapters to this effort, though they have presented their work in a manner accessible to a wide audience. By compiling within a single volume papers that share a focus on human settlement and biological diversity, we hope to provide readers with both a sense of the varying relationships between human populations and other species and a range of possible approaches to assessing problems in the human sciences of relevance to conservation biologists.

1.3 Status of the Field

Why is this field so underdeveloped? One possible answer lies in the expansive social, economic, and cultural variation found within the current human species and the complex changes occurring to populations of *Homo sapiens* within these domains. Because of this variation, researchers find it difficult to characterize trends in the relationship of a human settlement to communities of other biotic species without considering the array of production systems and their human-associated species, the access of its human inhabitants to economic assets and technology, and their attachment to a broader economy.

Another reason for this field's lack of development is its unique placement along the well-guarded boundary separating the biological sciences from the social sciences – a *no-man's land* infrequently traversed in academia that requires researchers to venture into areas beyond their area(s) of specialization. Several contributors to this volume are trained or experienced in both the biological and social sciences, and most conduct research programs either outside university settings or within academic institutes of applied analysis, where interdisciplinary traditions survive and are encouraged. The role of humans in biodiversity conservation is of interest to social scientists, and its problems could benefit from social scientific analyses – *Homo sapiens* is a primate of remarkable adaptive and cooperative abilities, with highly developed facilities for tool invention, production, and use and a unique capacity for communication. These are the foundations of *culture*, the means by which people adapt to, and transform, their natural and human environmental surroundings (Steward 1955). By developing and transmitting culture, humans have acquired behavioral flexibilities that free them from the constraints of biological adaptation which limit the distribution of other species. In so doing, our species has imposed its populations and their behaviors, and populations of associated species and their behaviors, on an unprecedented range of interrelated ecological systems.

1.4 Organization and Content

The volume is divided into three sections. The first comprises chapters presenting general theories and broad empirical relationships, which help explain dramatic changes in the patterns of occurrence of terrestrial and aquatic species that have developed in parallel with human population growth, migration, and settlement. The second section focuses on specific biomes and ecosystems as the context for human interactions with other species. Beyond their informational content, these chapters provide insights into the utility of using demographics and human geography to evaluate relationships between human settlement and the population dynamics of both native and nonnative species. The third and conclusive section comprises a discussion of the prior two sections by geographers Thomas Crawford and Deirdre Mageean.

The volume begins with a chapter by Lee Hachadoorian, Stuart Gaffin, and Robert Engelman, who present a map of the world's population future (Chap. 2). To generate a map of spatially distributed human population for the year 2025, the authors apply two simple, but well-considered, extrapolation techniques to project gridded population data. To test their extrapolation techniques, to identify systematic weaknesses in these methods, and to help them produce an algorithm that circumvents methodological pitfalls, they apply their techniques to a time series of spatial population estimates of the United States published by the US Census Bureau. Their research wrestles with basic problems of projection anomalies that need to be considered if demographers are to project human population's spatial distribution, whether globally, regionally, or locally, and it subdues these problems with reasonable, applicable solutions. The essay should prove extremely valuable to those undertaking similar challenges of spatially explicit projections of human population.

The essay by Christopher Small (Chap. 3) characterizes the global physiographic distribution of human population using overlays of Earth's physiographic features with that of temporally stable lights sensed at night from satellites. Small's objective is to provide a quantitative description of modern population distribution using a few basic environmental factors that may also influence the spatial distribution of biodiversity. His results suggest that because the vast majority of human population growth is projected to occur in and around urban areas, or in urbanizing areas, continued population growth in the biologically rich humid tropics and subtropics will likely focus on specific physical environments where urbanization is currently taking place. However, population remains stable or slowly growing in most of the rural areas of tropical and subtropical regions. Small's findings suggest that human adaptation to climate will cause expansion of the human habitat, whereas clustering with respect to the physical landscape will result in a simultaneous spatial concentration of population within the expanding habitat. The spatial distribution of population is strongly localized with respect to continental physiography (elevation, coastal, and fluvial proximity) but much less localized with respect to climatic parameters (annual mean and range of temperature and precipitation), resulting densities varying considerably.

In their chapter, Jeffrey McKee and Erica Chambers (Chap. 4) explore the role of human population in biodiversity loss, considering both total population and behavioral characteristics of population. Taking a global perspective, this study expands on McKee's previous arguments that sheer numbers of people and population density in the prehistoric and historic past has led to considerable biodiversity loss. Here the authors consider two variables that introduce characteristics of human behavior, addressing the important consideration that it is not necessarily just the number of people present but *what people do* that affects biodiversity. Their regression analyses of global population and biodiversity data, using information presented at the national level, indicate varying importance of gross national product and agriculture in determining the number of threatened species per nation, though ultimately preserving the importance of population density as a key factor in global biodiversity loss in modern settings as well as the past.

In Chap. 5, Richard Cincotta reviews three models that he considers relevant to human–biodiversity relationships and suggests that these could be modified and applied by researchers to help conceptualize, hypothesize, and in some cases, predict system dynamics. The first model, which applies the average adult body weight of *Homo sapiens* to statistical models predicting mammalian herbivore and predator density, suggests the degree to which our species has modified the landscape and channeled energy and nutrient flux in order to achieve densities that are today nearly three orders of magnitude greater than predicted for preagricultural humans. The second model examined by Cincotta makes explicit three types of risks to the viability of populations of native species: risks within the protected area, risks from between-area hazards, and risks at the reserve perimeter. The third model examined appears as a graphic representation of Boserup's (1965) theory that the demands of increasing human population drove agricultural innovation, in an attempt to make this theory more useful to ecologists.

Katalin Szlavecz, Paige Warren, and Steward Pickett review a very important, and much neglected, aspect of the human population–biodiversity interface, namely that existing in and near urban settings (Chap. 6). Although much of the focus on biodiversity, and indeed much of the focus of the chapters in this volume, involves more remote settings, an increasing amount of the Earth's surface is covered by dense human settlement and infrastructure best described as "urban." In their overview of biodiversity in such settings, Szlavecz and colleagues describe a variety of studies that document remarkable species richness in localities with high densities of human population. Several of the studies discussed explore how biodiversity changes with the ecological shifts accompanying dense concentrations of people in highly altered environments. As urban growth continues at unprecedented rates over the coming decades, the potential role of densely settled places in global biodiversity will grow markedly, as will the need to increase our understanding of this biodiversity.

Robin Abell, Michele Thieme, and Bernhard Lehner (Chap. 7) note that the literature on the use of landscape indicators to assess aquatic ecological integrity has grown substantially, but primarily through studies conducted in the developed, data-rich temperate world. In their chapter, they show how global and regional threat assessments can be undertaken, inclusive of data-poor regions, using human population density estimates as a "coarse proxy" for more specific indicators of disturbance. They warn, however, that although such proxies can correlate with biotic and abiotic measures of threats to freshwater species viability, they are best used for priority setting at the global, regional, and large-watershed scales. And they argue that such surrogates provide the conservation planner with neither an indication of the specific nature of local aquatic-ecological threats nor of possible policy solutions. Abell and her colleagues contend that research should address key data gaps that hinder scientists' ability to identify thresholds and ultimately mitigate threats to native aquatic populations, such as the relationships between local land use patterns and disturbances, and aquatic habitat quality. The authors conclude by noting a fortuitous shared interest: human communities' desire to safeguard freshwater quality overlaps significantly with the goal of bio-conservationists to maintain native aquatic communities.

The essay by Flora Lu and Richard Bilsborrow (Chap. 8) focuses on five distinct indigenous peoples in the Ecuadorian Amazon, examining how they differ demographically and how these demographic differences influence economic activities and, ultimately, environmental impacts. This sort of study, which examines humans as part of ecosystems composed of people and a range of other species, is of interest in its own right. The insights it provides on human adaptation – essentially on the *ecology* of people with respect to various demographic characteristics – are useful to bio-conservationists and reserve planners because it explores how the peoples in question use their resources and how varying patterns in resource use effect local biodiversity. To some, its greatest contribution may be in revealing some of the rich complexity of the ecology of indigenous peoples, emerging as contrasting approaches to survival in broadly similar environmental settings. As the authors state in their introductory comments, too often indigenous peoples are characterized as maintaining similar, predictable relationships with their natural surroundings. Results of this study point up considerable demographic and economic differences among the five groups considered, revealing often complicated relationships between human demographics and resource exploitation.

L.J. Gorenflo's chapter on the Apache Highlands (Chap. 9) incorporates a range of data on human demographics into conservation planning at an ecoregional scale. The essay focuses on the spatial distribution of population, describing this arrangement for two recent census years and exploring possible reasons among demographic processes for how these distributions came to be. In addition to this spatial perspective, the paper maintains a temporal perspective as it describes the prehistoric and historic roots of human settlement in the Apache Highlands, providing a sense of the sparse human settlement within this ecoregion, and the spatial distribution of species composition that developed. Through comparing recent human settlement with locations that are critical to the conservation of biological diversity, the study identifies a density threshold beyond which sites containing populations of key species rarely occur. Projections of human settlement for 2010 in both the Mexico and United States portions of the ecoregion identify potential conservation sites where this threshold will be exceeded, thereby likely facing human threats in the near future. Such studies, based largely on available census data, provide potential means of integrating basic types of human data into long-term conservation planning.

The paper by Li An, Marc Linderman, Guangming He, Zhiyun Ouyang, and Jianguo Liu (Chap. 10) explores the potential impacts of human settlement on biodiversity in the Wolong Nature Reserve in southwestern China. Noting that different underlying demographic mechanisms and economic conditions can have varying effects on both the total number of people living in an area as well as ultimate human impacts on the natural environment, this team of authors seeks to identify which mechanisms generate human impacts on a reserve that is essential to the maintenance of panda populations. Their approach involves agent-based computer modeling to simulate the role of various demographic and socioeconomic changes. Analyses conducted examine the roles of mortality, family planning (fertility and marriage age), mobility, and economic factors in determining total population and total households, both felt to have implications on the nature reserve, as well as direct impacts on panda habitat. Results of their computer simulations, presented in quantitative and cartographic form, reveal contrasting roles of variables on local demographics and panda habitat. Such increased understanding of how a range of human variables affect the human and natural environments provide a basis for designing programs that contribute to conservation goals, development goals, or both.

Working at a national scale in Madagascar, the study by L.J. Gorenflo, Catherine Corson, Kenneth Chomitz, Grady Harper, Miroslav Hónzak, and Berk Özler (Chap. 11) explores the importance of population as a predictor of adverse environmental impacts in the context of other possible socioeconomic, infrastructure, and physical indicators. The issue addressed is an important one, for characteristics of population – including level of demand and land use patterns – often are important considerations in assessing human impact. Results of the Madagascar study bear this out. The strongest predictor of deforestation between 1990 and 2000 was access, via roads and footpaths, with population itself a much less important determinant of where forest loss would occur. People certainly are responsible for cutting trees in Madagascar, but getting to the forests and moving timber and fuelwood to urban markets proved to be stronger predictors of deforestation than demographics alone. In this case, factors that influence transportation infrastructure provide a potential lever by which development policy can help to maintain forest habitat that is critical to the survival of a large number of endemic species.

In the book's final section, Thomas Crawford and Deirdre Mageean (Chap. 12) – both social scientists who have made substantial contributions to environmental research – look back at the National Research Council's (NRC) 2001 report entitled "Grand Challenges in Environmental Sciences," the objective of which was to identify the most salient challenges that multidisciplinary research could tackle during the twenty-first century, and to suggest, within each of these challenges, that represent likely avenues of progress. Two of the eight challenges are germane to this book – *biodiversity loss* and *land use and cover change*. After organizing the book's chapters based upon their fit into these themes, Crawford and Mageean discuss gaps, unresolved issues, and new approaches that should be considered from a social scientists' view.

1.5 Studying Human Population's Interactions with Biological Diversity

As the reader of this volume will discover, the relationships between our species and the populations of other species on Earth are, by no means, simple matters. No single variable among the set of human geographic and demographic indicators examined over the course of the chapters in this book appears to fit all the needs of researchers endeavoring to determine the relationships between humans and native species. Neither should researchers expect a single methodological approach to fit all situations when they set out to estimate the viability of native nonhuman populations in the vicinity of human settlement and activity or to predict the ecological outcomes of future human settlement.

Readers will also discover that seeking to understand current relationships between humans and other species, and predicting relationships in the future, can be quite demanding. We suggest that the challenges of understanding such relationships result from three sources of variation: the variation in native species' abilities to withstand the proximity of human settlement and associated nonnative species, the variation in the resilience of ecological systems to human-induced perturbation, and the variation of behaviors of human communities themselves. All vary greatly and are likely to vary even more in a future that includes dramatic climatic changes, technological innovation, and a larger, more consumptive human population.

Together, the authors of this volume have shown that select types of human geographic and demographic data can be of substantive value in research efforts focused on assessing biodiversity-related ecological trends. The reasons are straightforward and are borne out repeatedly in various ways throughout the volume. The first lesson is that human density has a powerful influence – directly or indirectly – on the viability of populations in the vast majority of native species. Second, and somewhat inimical to this message, is that changes in human population density are not necessary conditions for dramatic anthropogenic ecosystem change. Various forms of human activity and deposition of products of human activity, from within or beyond the apparent boundaries of ecosystems, have been sufficient to modify the cycling of nutrients and energy through native ecosystems, to degrade habitat quality for certain species, and ultimately to alter native species composition. And other human activities, particularly some associated with human infrastructural development (such as road building and dam construction), have accelerated or exacerbated losses of biological diversity in regions where human settlement was initially sparse. The third lesson is that scale matters. Several authors conclude that indicators of human distribution or settlement qualities provide acceptable indications of biodiversity-relevant ecosystem qualities at large-scales (global, regional and river-basin levels) particularly when other, more proximate indicators remain unavailable or incomparable. However, although such research warns of the significant challenges that human population growth, settlement, and urbanization present to the bio-conservationist intent on maintaining the components of Earth's biological diversity that remain intact, it does not, by itself, provide insights into the factors associated with settlement and land use that may be amenable to proactive policies and management.

Although the concerns of this volume's essays revolve around problems associated with conserving biological diversity, at the core are the dynamics of our own species. It is humbling to acknowledge that, at the time of the writing of this volume, more human offspring are born every day (around 350,000, on average) than the total living individuals in all of the other species of great apes (family: Hominidae) combined (Cincotta and Engelman 2000) – more than the world's combined populations of our closest biological relatives, chimpanzees (*Pan troglodytes*), bonobos (*P. paniscus*), gorillas (*Gorilla gorilla*), and orangutans (*Pongo pygmaeus*). Such a statistic suggests not only the power and dominion of modern *Homo sapiens* but also the magnitude of present and future challenges facing those involved in biodiversity conservation.

References

Boserup E (1965) The conditions of agricultural growth: the economics of agrarian change under population pressure. Aldine, Chicago

Cincotta RP, Engelman R (2000) Nature's place: human population and the future of biological diversity. Population Action International, Washington, DC

Cincotta RP, Wisnewski J, Engelman R (2000) Human population in the biodiversity hotspots. Nature 404:990–992

Gorenflo LJ (2002) The evaluation of human population in conservation planning: an example from the Sonoran Desert Ecoregion. Publications for Capacity Building, The Nature Conservancy, Arlington, Virginia

Hoare R, Du Toit J (1999) Coexistence between people and elephants in African savannas. Conserv Biol 15(3):633–639

Holdaway RN, Jacomb C (2000) Rapid extinction of the moas (Aves: Dinornithiformes): model, test, and implications. Science 287:2250–2254

Martin PS (1984) Prehistoric overkill: the global model. In: Martin PS, Klein RJ (eds) Quaternary extinctions: a prehistoric revolution. University of Arizona Press, Tucson, Ariz, pp 354–403

McKee JK, Sciullia PW, Foocea CD, Waite TA (2003) Forecasting global biodiversity threats associated with human population growth. Biol Conserv 115:161–164

Steward JH (1955) Theory of culture change: the methodology of multilinear evolution. University of Illinois Press, Urbana, Illinois

UNDP (United Nations Development Programme), UNEP (United Nations Environment Programme), World Bank, World Resources Institute (2000) A guide to world resources 2000–2001, People and ecosystems: the fraying web of life. World Resources Institute, Washington, DC

United Nations Population Division (2009) World Population Prospects, the 2008 Revision. United Nations, New York

Vitousek PM, Mooney HA, Lubchenco J, Melillo JM (1997) Human domination of earth's ecosystems. Science 277:494–499

Part I
General Theoretical and Empirical Considerations

Chapter 2
Projecting a Gridded Population of the World Using Ratio Methods of Trend Extrapolation

Lee Hachadoorian, Stuart R. Gaffin, and Robert Engelman

2.1 Introduction

Geographic information systems (GIS) are computer software used to describe and analyze spatial data (DeMers 2000). Through linking geographic locations with data that describe those locations, GIS provides a basis for examining a range of questions associated with the arrangement of variables in geographic space. Since human beings necessarily inhabit geographic space, population data are inherently spatial. GIS enables explicit investigations of the geography of human population, complementing the range of demographic methods used to measure other population characteristics.

Smith et al. (2001: 365–366) describe recent uses of GIS for demographic research, concluding that this tool makes demographic trend analysis possible for very small spatial areas. Such a capability could open up new possibilities for considering some of the likely interactions of present and projected human populations with nonhuman species. Most such species must retreat or, at a minimum, alter their behavior as human beings not only expand their occupation of physical space but also co-opt natural resources and whole ecosystems for human purposes and activities. Smith et al. nonetheless caution that "... GIS-based projection methods are complex and expensive. They require a substantial investment of time and effort ...." In the following paper, we show that it is feasible to use GIS to apply simple projection methods to a large number of small area projections. And, while GIS software and its application are not inexpensive, there is almost certainly no less expensive way to create millions of small area projections at once.

2.2 Recent Population Density Maps and Prior Projections Using Simple Methods

For interdisciplinary analysts, maps of current world population density are fundamental to many environmental and socioeconomic studies. To cite just a few diverse applications, such maps have facilitated demographic analysis of the earth's biodiversity hotspots (Cincotta et al. 2000), economic studies on the geography of

R.P. Cincotta and L.J. Gorenflo (eds.), *Human Population: Its Influences on Biological Diversity*, Ecological Studies 214,
DOI 10.1007/978-3-642-16707-2_2, © Springer-Verlag Berlin Heidelberg 2011

poverty and wealth (Sachs et al. 2001), and greenhouse gas emission inventories used in global climate change simulations (Van Aardenne et al. 2001; Graedel et al. 1993).

Gridded population density maps have been developed at regional and global scales. Methods and data have included both census enumerations (Tobler et al. 1995; CIESIN 2000, 2005a) and modeled allocations of population data to smaller geographic units using supplemental information such as proximity to roads and remotely sensed patterns of night lights (ORNL 2002). However, little theoretical work has been applied to the problem of developing projected versions of such maps, despite the fact that projected versions would likely be of similar interest to the users of present day maps (CIESIN 2005b).[1] One reason for the limited work on projected grids is the obvious challenge of making, or obtaining, demographic projections at small spatial scales for many world regions.

The United Nations routinely develops 50-year projections for the world's approximately 230 countries (United Nations Population Division 2003). These projections rely on state-of-the-art cohort-component modeling that is the commonly accepted tool for world and national projections. In addition, the United Nations also makes projections to 2030 for each country's urban and rural populations and to 2015 for 30 of the largest urban agglomerations with 750,000 or more inhabitants. These subnational projections are presented as numerical totals for each country and not as spatial data. In addition, they are based on logarithmic extrapolation of changes in the urban–rural population ratio for each country, not the cohort-component methods used in the national projections (United Nations Population Division 2003).

In contrast, present day population maps exist at spatial scales as small as 0.5 min latitude and longitude (~0.9 km at the equator) (ORNL 2002; CIESIN 2004) or 2.5 min latitude and longitude (~4.6 km at the equator) (CIESIN 2000, 2005a). At the latter scale, the maps include density estimates for almost nine million grid cells over the Earth's land areas. It clearly is not feasible to apply age-cohort modeling to such a vast array of grid cells as it would require assembling vital rate information (fertility, mortality, and migration rates) for each grid cell. Such data are not available, and without them, standard age-cohort projections are not an option.

But even if cohort-component modeling on such scales were technically possible – in a sense the ultimate spatial disaggregation of world population projections – an important question is whether the accuracy of the resulting projections would be worth the vast effort they required in comparison to simpler methods. Studies of the relative accuracy of simpler versus complex projection methods suggest that the

[1]Public and educational interest in such projections is evidently high, judging by the popularity of a short film evidently based on a simple artistic rendering of population change that has circulated since the 1970s (original version distributed in 1972) and been revised and reissued in 1990 and 2000 (latest version, see Population Connection 2000). The film makes its point by animating the growth of world population as an extensification and intensification of dots across the surface of the earth.

complex methods may perform no better than simple methods – with respect to forecasts of *total* population – when retrospective analyses of accuracy have been done (Smith 1997; Long 1995; White 1954). In his survey of published ex post comparisons of projection accuracy from simpler methods as compared to more complex causal and cohort models, Smith (1997: 560–561) concluded that "... There is a substantial body of evidence ... supporting the conclusion that more complex [population projection] models generally do not lead to more accurate forecasts of total population than can be achieved with simpler models ...." Alternative viewpoints on this question are made in Beaumont and Isserman (1987), Ahlburg (1995), and Keyfitz (1981).

We are aware of only three other published gridded world population projections using simple methods. In the first case, Gaffin et al. (2004) attempted to model a 2025 GPW projection using a "constant-share" method. This method requires a launch year (1990) and a projection (derived from the 1996 Revision of the United Nation's *World Population Prospects*). The share of national population in the grid cell is calculated for the launch year and held constant for any future projection. If a grid cell holds 1% of national population in 1990, it is assigned 1% of the projected 2025 national population. This forces all small area trends to conform to the large area trends, so subareas that are losing population prior to the launch year will be seen to gain population thereafter (provided that the large area is gaining population). This kind of trend reversal was noted as problematic and led to the current attempt by the authors to apply more sophisticated ratio methods. Note, however, that Smith et al. (2001) still define the methods we apply in this chapter as "simple" projections methods compared to other trend methods (such as ARIMA time series models).

Thornton et al. (2002) work with historical data at the administrative level within developing countries. They used 1980–1990 administrative unit population growth rates and held these rates constant throughout the forecast period to 2050. Gridding of the administrative areas was then applied using a smoothing algorithm. Their method does not ensure consistency with the United Nations national projections for the countries they study, and discrepancies with the UN 2050 projections arise. In cases where these discrepancies were large, the results were rescaled to match the UN figures.

Finally, the Center for International Earth Science Information Network (CIESIN 2005b) produced future estimates of population out to 2015, released in conjunction with their Gridded Population of the World Version 3 (CIESIN 2005a). These estimates are made based on logarithmic extrapolation of historical data at the administrative level within each country. For some countries, administrative boundaries are proprietary and not available for release. A scale factor was applied to all subareas to force compliance with the UN national projections. Population was then distributed over the grid using a proportional allocation method.

In contrast with Thornton et al. (2002) and CIESIN (2005b), we project population at the grid cell level. Our projections combine two different extrapolation methods so as to reduce anomalous results, and we also build agreement with the UN projections into the algorithms.

2.3 Methods

Trend extrapolation methods, long ignored in favor of cohort-component methods, have made a comeback in recent years due to theoretical advances and their relatively low cost and data requirements (Smith et al. 2001). Ratio methods rely on the existence of independent projections of larger areas. While most small-area projections use states or other administrative boundaries as the subarea of a larger administrative unit, in this chapter, we treat the Gridded Population of the World Version 2 grid cells (described below) as subareas of countries.

To carry out the Geographic information systems (GIS) calculations, we used ArcView 8.3 (ESRI 2002) with ESRI's Spatial Analyst extension to conduct analysis of raster grids. Spatial Analyst allows the user to perform various logical and mathematical operations on grids. Among the simplest operations is arithmetic. Addition, for example, will generate an output grid where each grid cell is assigned a value equal to the sum of the values of the corresponding cells from each input grid (that is, those cells which are in the same spatial location). The result is much the same as creating a spreadsheet whose A1 cell is equal to the sum of the A1 cells from several other spreadsheets.

For our base year maps, we work with the 1990 and 1995 Gridded Population of the World, Version 2, databases (GPWv2), maintained and disseminated by CIESIN at Columbia University. These data are partly historical and partly estimated. For example, the United States conducted censuses in 1990 and 2000, so the 1990 data in GPWv2 represent actual census counts, while the 1995 data represent a population estimate between census years.

The GPWv2 and GPWv3 datasets are available at various resolutions down to 2.5 min, or ~4.6 km at the equator (becoming slightly smaller towards the poles). The population is allocated in proportion to the area of each grid cell, with those grid cells crossing administrative boundaries being assigned an area-proportionate population from each administrative unit. Year 2000 data are currently available as part of GPW3, although we rely on 1990 and 1995 data from GPWv2, available at the time this analysis was conducted.

The two ratio methods that we apply require two historical data points: a base year (1990) and a launch year (1995). The first is referred to as the "*shift-share*" method by Smith et al. (2001). With this, one calculates the annual rate of change in the *subarea fractional share of national population* between the base year and the launch year:

$$f_{\text{subarea}}(1995) = \frac{\left[\dfrac{P_{\text{subarea}}(1995)}{P_{\text{national}}(1995)} - \dfrac{P_{\text{subarea}}(1990)}{P_{\text{national}}(1990)}\right]}{1995 - 1990}. \tag{2.1}$$

This trend factor is held fixed and extrapolated to the projection year – 2025 for this study – to yield an extrapolated share. The extrapolated share is then applied to the large area projection for 2025 to generate the small area projection for 2025:

$$P_{\text{subarea}}(2025) = P_{\text{national}}(2025) \cdot \left[\frac{P_{\text{subarea}}(1995)}{P_{\text{national}}(1995)} + f_{\text{subarea}}(1995) \times [2025 - 1995]\right]. \tag{2.2}$$

This method is known to suffer face validity problems (Smith et al. 2001) such as trend reversals and negative population projections. These problems did occur with the present projection and are discussed in the results section.

While we use the simplest shift-share method of linear extrapolation from two data points, Gabbour (1993) describes other shift-share methods that use more than two data points and nonlinear extrapolation methods (such as exponential and logistic). Since more than two data points are used, he calculates a Pearson correlation coefficient for each subarea and retains the method that produces the highest R^2. These methods remain as possibilities for future investigation.

The second method we employ is referred to as the "*share-of-growth*" method (Smith et al. 2001). This also requires a base year (1990) and a launch year (1995). We calculate the proportion of national growth that occurs in each subarea:

$$f_{\text{subarea}}(1995) = \left[\frac{P_{\text{subarea}}(1995) - P_{\text{subarea}}(1990)}{P_{\text{national}}(1995) - P_{\text{national}}(1990)}\right]. \tag{2.3}$$

This fraction is then multiplied by the projected change in national population between the launch year and the projection year and added to the launch year subarea population to obtain the future subarea population projection.

$$P_{\text{subarea}}(2025) = P_{\text{subarea}}(1995) + f_{\text{subarea}}(1995) \times [P_{\text{national}}(2025) - P_{\text{national}}(1995)]. \tag{2.4}$$

Unlike the shift-share method, trend reversals will only occur with the share-of-growth method if a trend reversal is projected for the national area by the United Nations projections.

Figure 2.1 depicts the four generic trend reversal cases that can occur with this method. In Fig. 2.1a, for example, both the nation and subarea have growing population during the base period, and the UN projects a declining national population during the forecast period. The share-of-growth model will forecast a declining population for the subarea during projection. Figure 2.1a, b corresponds to the situation when f_{subarea} in (2.3) is positive, and Fig. 2.1c, d corresponds to a negative value for f_{subarea}.

2.4 Results

Comparisons of the shift-share and share-of-growth models with cohort-component projections lead us to strongly prefer share-of-growth. We discuss the results of this comparison, followed by problems specific to each model. We address in detail

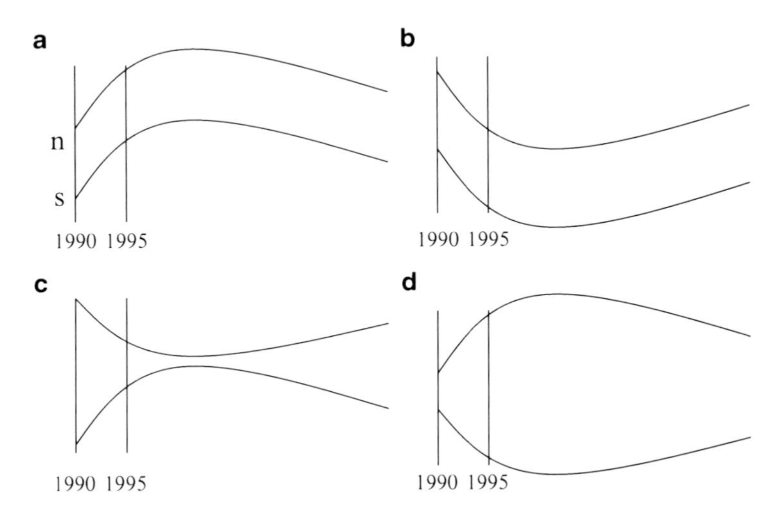

Fig. 2.1 Four generic types of trend reversals that can occur with the share-of-growth method (*n* national population; *s* subarea population)

those circumstances under which share-of-growth fails to perform as well. Finally, we present our criteria for choosing between these models in creating a combined-model gridded population projection.

2.4.1 Comparison of the Simple Projections with US Census Bureau State Projections

We assess the shift-share and share-of-growth methods against US Census Bureau state-level projections (US Census Bureau 2005). The gridded projections for the United States are redone using all three ratio methods described (constant-share, shift-share, and share-of-growth) using the GPWv2 1990 and 1995 population grids and using the sum of Census Bureau state-level projections as the large area projection for the United States (which differs from the United Nations projection for the country). The grid cells are then aggregated by state to generate a total 2025 population for each state. We then assess forecast "accuracy" by treating the Census Bureau projections as "observed data." This is, therefore, not a genuine measure of accuracy. Rather, it is a measure of how these ratio methods compare to standard cohort-component methods. We do not assess bias; since ratio methods are designed to conform to an independent large area projection, state-level errors will cancel when summed, and bias is necessarily zero.

Comparisons to the Census Bureau projections were made using standard mean absolute percentage error (MAPE) and geometric mean absolute percentage error (GMAPE). The mean is a reasonable measure of central tendency for a normally distributed variable, while the geometric mean is a better measure of central tendency for a variable whose *logarithm* is normally distributed. One way to determine whether a variable is normally distributed is to examine its quantile–normal plot, which is a plot of the variable against a normally distributed theoretical variable having the same mean and standard deviation (Hamilton 1992). If the quantile-normal plot of a variable falls on a straight line, that variable is normally distributed.

A variable whose logarithm is normally distributed is referred to as log-normally distributed. Log-normal distribution is typical of left-bounded, right-skewed variables such as absolute percentage errors (APEs). The log-normal plot of a variable is a plot of the natural log of a variable against a theoretical variable whose natural log is normally distributed and has the same mean and standard deviation. The quantile-normal plots for the share-of-growth APEs and the shift-share APEs do not fall on a straight line, calling into question the usefulness of using MAPE as a comparison of accuracy. The log-normal plot of the share-of-growth and shift-share APEs are closer to a straight line, justifying the reporting of GMAPE here.

We also note that high outliers (which occurred in our case for small states such as Nevada and the District of Columbia) suggest the utility of transforming the APEs. Emerson and Stoto suggest a maximum of 20 for the ratio of the highest to lowest APE (Emerson and Stoto 1983, cited in Swanson et al. 2000). For all three models we consider, the ratio of the highest to lowest APE is larger than 20. For example, for the constant-share model, the highest APE (24.90% for West Virginia) is approximately 138 times as large as the lowest APE (0.18% for Maryland). For the shift-share and share-of-growth models, this ratio is 267 and 103, respectively. We therefore also report our comparisons in MAPE-R (Table 2.1). This measure of accuracy applies the modified Box–Cox transformation described by Swanson et al. (2000) and then uses regression to re-express the transformed APEs in the same scale as the original APEs.

In conforming to Census Bureau cohort-component projections, the share-of-growth projection method performs noticeably better than either of the other ratio methods. In comparing shift-share with constant-share, we do not feel that the MAPE metric is appropriate for reasons cited above, and we do not feel that the GMAPE metric is appropriate because the APEs produced by the constant-share model are not log-normally distributed. We therefore use the MAPE-R metric for comparison purposes, and by this measure, shift-share appears to perform slightly better than constant-share (9.18% as opposed to 9.29%).

Table 2.1 Absolute percentage error differences between the three ratio method 2025 projections for the U.S. states and the Census Bureau 2025 projections

Model	Absolute percentage error metric		
	MAPE (%)	GMAPE (%)	MAPE-R (%)
Constant-share	10.12	7.33	9.29
Shift-share	13.80	8.05	9.18
Share-of-growth	7.96	5.15	5.49

2.4.2 Problems of the Shift-Share Method

The shift-share method suffers face validity problems. As with the constant-share method used by Gaffin et al. (2004), it also produces trend reversals. A subarea that is growing, but not growing as fast as the national population, will have a negative trend in its population share. For short-horizon projections, this will likely not be a problem; but for long-horizon projections, the trend can produce a ratio small enough to show a population loss in spite of the fact that both the subarea and national area show population increases between the base and launch years. Additionally, this method is known to perform poorly for areas losing population. Since many of the GPW grid cells losing population are rural areas that already have low populations (and therefore a very low share of national population), this method can lead to projections of negative population shares (and therefore negative population, an obvious impossibility in human terms). Figure 2.2 shows areas of the world that contained some negative grid cells using shift-share.

2.4.3 Problems of the Share-of-Growth Method

Negative populations are also possible with the share-of-growth method (Fig. 2.3), though not as widespread as with the shift-share method (62 countries with *any* negative grid cells using share-of-growth, versus 89 such countries with the shift-share method). On the other hand, the share-of-growth method generated grid cells with very large negative values for countries experiencing certain kinds of demographic trends. This is best explained in relation to the issue of trend reversals, summarized in Fig. 2.1.

Reversals like Fig. 2.1a, b are plausible on their face as the subarea is experiencing the same trend reversal as the nation. However, reversals such as Fig. 2.1c, d, without further demographic information on the region, are anomalous.

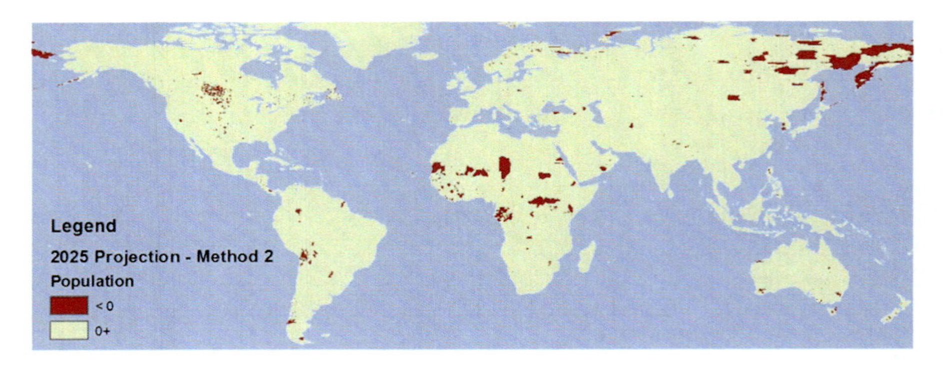

Fig. 2.2 Areas with negative 2025 projected population densities using the shift-share method

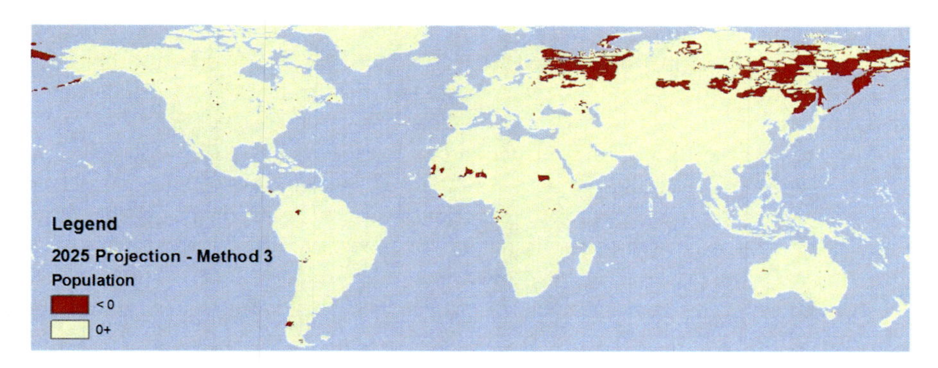

Fig. 2.3 Areas with negative 2025 projected population densities using the share-of-growth method

Portugal exemplifies the trend reversal danger of the share-of-growth model. Portugal experienced urban growth and rural decline between the base year and the launch year, with an overall population decrease of 0.08% (the United Nations now estimates that Portugal's population grew from 1990 to 1995; we persist in using the 2002 Revision because it provides a useful example of this pitfall of the share-of-growth method). Then, from the launch year to the target year, Portugal is expected to gain by 2.22%. Rural areas in Portugal correspond to Fig. 2.1b and urban areas to Fig. 2.1c. Since the previous growth was in the urban areas, but the national growth undergoes a sign change, the share-of-growth method reverses the trend for urban areas and increases the population in rural areas. These increases in a large number of rural grid cells are offset in the ratio method by decreases in a small number of urban grid cells, with the end result that this method projects negative populations in the urban grid cells. The sum of all negative grid cells comes to 24.84% of the projected national population. Furthermore, zeroing out and renormalizing in this case would produce an absurd distribution of population, with zero population in formerly urban areas.

The most spectacular failure of the share-of-growth method did not, however, occur for a country experiencing trend reversal. It occurred for Russia, where the rate of change was much larger in the projection period than the base period and the country was experiencing decline of the national population in both periods. During the 30-year projection period, Russia was projected to lose 131 times the population as was lost during the 5-year base period, for an annual rate of loss 21.8 times faster during the projection period. The next largest problem countries were Lithuania (with 10.3 times the rate of loss) and Ukraine (with 3.8 times the rate of loss).

2.4.4 Combined Model Gridded Projection

As mentioned in the section on Census Bureau comparison, the share-of-growth projection method performs noticeably better than either of the other ratio methods.

In deciding which method to retain for each country in our world projection, we selected the share-of-growth method for this reason, as well as because the shift-share seems more susceptible to negative population projections. We rule out share-of-growth for those countries where population change reverses direction.

Although we performed these calculations for the entire world at once, each country's grid cells are independently trended and normalized to that country's national (historical and projected) population. Examining the results of the shift-share and share-of-growth methods for each country confirmed that for the vast majority of countries, the share-of-growth model had the best face validity.

As mentioned above, both methods can produce negative grid cells under the wrong conditions. The negative grid cells in each country were summed and expressed as a percentage of the projected national population. Particularly poor performers are listed in Table 2.2. One of the criteria used to determine the best projection was which method projected the smallest percentage of negative population. In general, a country that performed poorly with one model performed reasonably well with the other one. For many countries, we were able to eliminate the negative grid cell problem entirely by choosing between the models. Nonetheless,

Table 2.2 Countries for which at least one projection method resulted in negative person counts totaling greater than 5% of the projected national population

Country	Negative population as a percentage of national population, by method	
	Shift-share (%)	Share-of-growth (%)
Russian Federation	2	256
Gabon	28	3
Portugal	0	25
Mauritania	9	2
Bolivia	7	2
Lithuania	0	9
Gambia	9	0
Mongolia	6	0

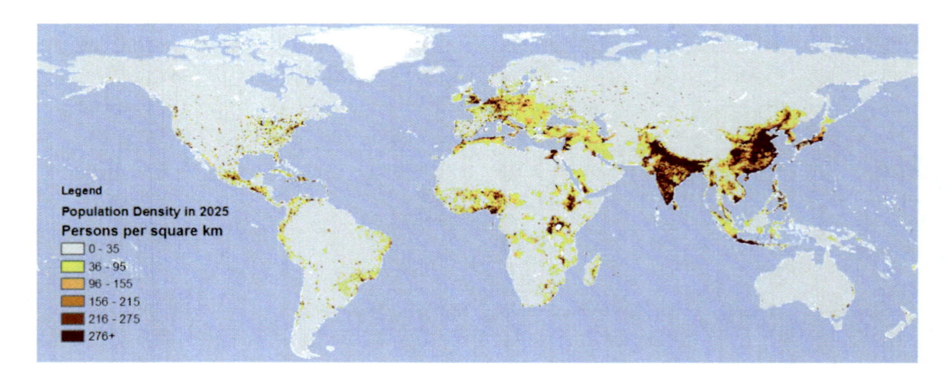

Fig. 2.4 Combined model using share-of-growth method for most counties and, to minimize anomalies, the shift-share for the remainder

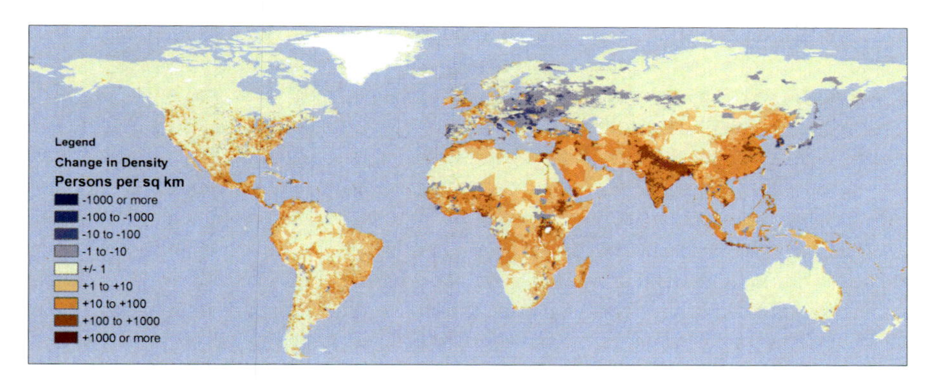

Fig. 2.5 1995–2025 population density change map

the combined model grid still had some countries with negative grid cells – that is, where both methods projected a negative population for some grid cells for a given country – though nothing larger than 2.69% (Gabon, which appears in Table 2.2 rounded to 3%). These remaining negative cells were set to zero and the grid was renormalized to the United Nations national projections. The final combined map for 2025 is shown in Fig. 2.4. A 1995–2025 change map for the combined model is shown in Fig. 2.5.

2.5 Conclusions

We have described the use of Geographic information systems (GIS) for small-area projections using ratio trend extrapolation methods and have commented more generally on the performance of these ratio methods. Our end result is a gridded world population projection. A wall chart based on this gridded projection is available as a PDF download at http://ccsr.columbia.edu/?id=research_population. It is important to bear in mind that all population projections are conditional forecasts, based on assumptions made about future trends, and these maps are no exceptions. In particular, mapped results are sensitive to the population changes that occurred in the reference period (in this case 1990–1995). For the same reason, we are not able to, at this time, integrate known changes in population distribution that occurred after the reference period.

This projection can be improved and extended in a number of ways. First, we rely on general considerations of face validity in choosing which method to apply to each country. A demographer closely familiar with a particular country may have very good reasons for criticizing our choice. In fact, we cannot rule out the possibility that the simple constant-share model may produce the most plausible projection for a given country. Without applying any new methods, the projection could be improved by being examined by demographers with expertise in each country. Indeed, with sufficient financial and other support, the projections could become the vehicle for a dialog with such demographers. This in turn could be expanded into further explorations of human–biodiversity interactions around the world.

Second, the projection can be improved by using high-quality subnational projections, especially for countries covering large areas, such as Russia, Canada, the United States, China, Brazil, and Australia. Those countries for which subnational projections are not available can at least be split into urban and rural areas so that United Nations urban–rural projections can be applied. As part of the GPW3 beta test, CIESIN (2004) has released a GPW-compatible grid of urban extents developed under the Global Urban–Rural Mapping Project (GRUMP). Using this urban extents grid, urban and rural populations can be trended and normalized independently.

Third, alternative trend extrapolation methods can be explored. When GPW has matured to the point where several grids are available, regression-based methods such as ordinary least squares, logistic, and ARIMA models can be used.

Finally, of course, as the GPW database continues to be updated, the reference period will lengthen and capture a longer time span, helping to average out more episodic changes in population that may occur in a 5-year or 10-year period.

Despite the uncertainties in mapping population projections – uncertainties which apply to the projections themselves – such exercises can tell us much about the likely future of population distribution worldwide, and hence, perhaps, the future of biodiversity as well. Seeing well-defined geographic areas of projected population decline in Africa, sub-Saharan, for example, serves as a reminder that urbanization powerfully influences rural population density despite high natural increase in such areas. The maps make clear that population change is anything but uniform spatially, and demographic diversity is characteristic of nations as well as of the world as a whole. Finally, such maps provide a powerful visual reminder that, despite large areas of projected population decline, the world overall will have many more people in 2025 than it does today. For the vast majority of nonhuman species, absent surprises or sustained efforts that slow human fertility more than the medium projection anticipates, it is likely to be a less accommodating planet.

References

Ahlburg DA (1995) Simple versus complex models: evaluation accuracy and combining. Math Popul Stud 5:281–290

Beaumont PM, Isserman AM (1987) Comment. J Am Stat Assoc 82:1004–1009

CIESIN (Center for International Earth Science Information Network) (2000) Gridded population of the world (GPW) version 2. Palisades Columbia University New York. http://sedac.ciesin.columbia.edu/gpw-v2/index.html?main.html&2. Accessed August 2004

CIESIN (Center for International Earth Science Information Network) (2004) Global rural-urban mapping project (GRUMP) alpha version. Socioeconomic Data and Applications Center (SEDAC) Columbia University Palisades New York. http://sedac.ciesin.columbia.edu/gpw. Accessed August 2006

CIESIN (Center for International Earth Science Information Network) (2005a) Gridded population of the world version 3 (GPWv3). Socioeconomic Data and Applications Center (SEDAC) Columbia University Palisades New York. http://sedac.ciesin.columbia.edu/gpw. Accessed August 2006

CIESIN (Center for International Earth Science Information Network) (2005b) Gridded population of the world: future estimates. Socioeconomic Data and Applications Center (SEDAC) Columbia University Palisades New York. http://sedac.ciesin.columbia.edu/gpw. Accessed August 2006

Cincotta RP, Wisnewski J, Engelman R (2000) Human population in the biodiversity hotspots. Nature 404:990–992

DeMers MN (2000) Fundamentals of geographic information systems, 2nd edn. Wiley, New York

Emerson J, Stoto M (1983) Transforming data. In: Hoaglin D, Mosteller F, Tukey J (eds) Understanding robust and exploratory data analysis. Wiley, New York, pp 97–128

ESRI (Environmental Systems Research Institute) (2002) ESRI GIS and mapping. www.esri.com

Gabbour I (1993) SPOP: small-area population projection. In: Klosterman R, Brail R, Bossard E (eds) Spreadsheet models for urban and regional analysis. Center for Urban Policy Research, Rutgers University, New Brunswick, New Jersey, pp 69–84

Gaffin SR, Rosenzweig C, Xing X, Yetman G (2004) Downscaling and geo-spatial gridding of socio-economic scenarios from the IPCC Special Report on Emissions Scenarios. Global Environ Change 14:105–123

Gaffin SR, Hachadoorian L, Engelman R (2006) Mapping the future of population (wallchart). Population Action International, Washington DC

Graedel TE, Bates TS, Bouwman AF, Cunnold D, Dignon J, Fung I, Jacob DJ, Lamb BK, Logan JA, Marland G, Middleton P, Pacyna JM, Placet M, Veldt C (1993) A compilation of inventories of emissions to the atmosphere. Global Biogeochem Cycles 7:1–26

Hamilton LC (1992) Regression with graphics: A second course in applied statistics. Wadsworth, Belmont, CA

Keyfitz N (1981) The limits of population forecasting. Popul Dev Rev 7:579–593

Long JF (1995) Complexity accuracy and utility of official population projections. Math Popul Stud 5:203–216

ORNL (Oak Ridge National Laboratory) (2002) Landscan. http://www.ornl.gov/sci/gist/landscan. Accessed August 2004

Population Connection (2000) World population: a graphic simulation of the history of human population growth (video), Washington DC. www.populationeducation.org/index.php?option=com_content&task=view&id=175&Itemid=4

Sachs JD, Mellinger AD, Gallup JL (2001) The geography of poverty and wealth. Sci Am 285:70–75

Smith SK (1997) Further thoughts on simplicity and complexity in population projection models. Int J Forecast 13:557–565

Smith SK, Tayman J, Swanson DA (2001) State and local population projections: methodology and analysis. Kluwer Academic, New York

Swanson DA, Tayman J, Barr CF (2000) A note on the measurement of accuracy for subnational demographic estimates. Demography 37:193–201

Thornton PK, Kruska RL, Henninger N, Kristjanson PM, Reid RS, Atieno F, Odero AN, Ndegwa T (2002) Mapping poverty and livestock in the developing world. International Livestock Research Institute, Nairobi

Tobler WR, Deichmann U, Gottsegen J, Maloy K (1995) The global demography project, technical report no 95-6. National Center for Geographic Information and Analysis, University of California, Santa Barbara

United Nations Population Division (2003) World urbanization prospects: the 2003 revision, data tables and highlights. United Nations, New York

US Census Bureau (2005) State population projections. http://www.census.gov/population/www/projections/stproj.html

Van Aardenne JA, Dentener FJ, Olivier J, Klein GJ, Goldewijk CGM, Lelieveld J (2001) A 1 $\times$ 1 resolution data set of historical anthropogenic trace gas emissions for the period 1890-1990. Global Biogeochem Cycles 15:909–928

White HR (1954) Empirical study of the accuracy of selected methods of projecting state population. J Am Stat Assoc 49:480–498

Chapter 3
The Human Habitat

Christopher Small

3.1 Introduction

Studies of global ecology and biodiversity conclude that there is now no ecosystem on Earth that is not impacted by human population (Vitousek et al. 1997) and that human populations are impacting the ecosystems that host a disproportionate fraction of Earth's biodiversity (Cincotta et al. 2000). The degree to which this will change in the future depends on present and future distribution of humans and the impact they have on surrounding ecosystems. Human adaptation and habitat expansion during the past 40,000 years must be viewed as an evolutionary success for humans, but the ecological impacts of the resultant population growth are not yet known (see Turner et al. 1990 for an extensive overview). The global population growth rate is currently decreasing but demographic momentum implies that population growth will continue according to most forecasts until at least the year 2100 (O'Neill and Balk 2001). At the same time, widespread urbanization has the effect of concentrating the growing population in dense settlements at unprecedented rates (United Nations Population Division 2002). The ecological impact of human population movement and settlement is not confined to population centers, rather the nature of the spatial distribution of population affects the environment both where the population resides and elsewhere. For this reason, understanding the spatial and environmental consistencies in the distribution of human population distribution is critical to understanding the impacts that population growth could have on the rest of the Earth system in the future.

The population of Earth's landmasses exhibits spatial structure at a wide range of scales in spite of the fact that human settlement is not a centrally planned activity at global scales (Small et al. 2010). Many of these patterns have yet to be systematically quantified and explained on a global scale. The potential significance of geographically influenced spatial scaling in population distribution bears directly on questions of human-induced changes in the environment (Cohen 1997) and has numerous socioeconomic implications (e.g., Sachs 1997; Gallup et al. 1999). This chapter discusses relationships between the spatial distribution of population and some of the geophysical forces commonly assumed to determine the human habitat. The objective is not to provide a causal explanation for population distribution but

R.P. Cincotta and L.J. Gorenflo (eds.), *Human Population: Its Influences on Biological Diversity*, Ecological Studies 214,
DOI 10.1007/978-3-642-16707-2_3,

to provide a systematic, quantitative description of modern population distribution with respect to some basic environmental factors, which may also influence the spatial distribution of biodiversity. The methodology used here could be extended to include a larger number of environmental, demographic, political and socioeconomic variables to quantify and better understand population distributions and their relationships to these factors.

Human populations are not uniformly distributed on Earth's landmasses. The spatial distribution of the global human population at any time shows large variations over a wide range of spatial scales (Fig. 3.1). Understanding this distribution is

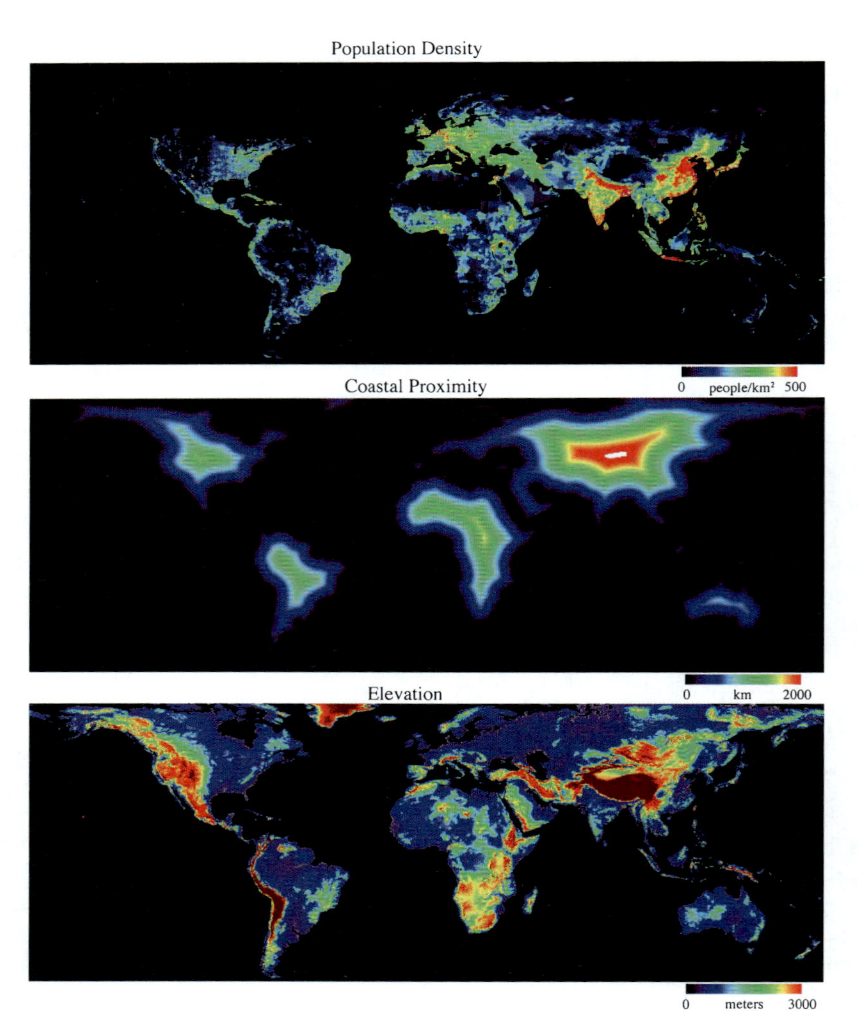

Fig. 3.1 Global distributions of (**a**) population (Small 2004), (**b**) coastal proximity and (**c**) elevation. Note that population density is shown on a Log_{10} scale to accommodate the wide range

fundamental to understanding the relationships between humans and the environment. The study of human demography has made great advances in understanding the structure and growth of human populations but has placed relatively little emphasis on the spatial dimension of human distributions. Livi-Bacci (1997) states: "In order to study long-term demographic growth, we must take 'space' into account and all that it implies, in particular land, the products of the land (food, manufactured goods, energy) and those characteristics which determine settlement patterns. Demography has for too long ignored or, at best, paid scant attention to these themes and so deprived itself of valuable interpretative tools." The spatial dimension is fundamental to both the Malthusian and Boserupian extremes of the population debate (Malthus 1798; Boserup 1965, 1981) and yet it remains one of the least understood or studied. Similarly, the study of geophysical processes has, until recently, failed to address the spatial relationship between Earth's physical and human systems. Ecologists and geographers have considered the spatial dimensions of these interactions but have been constrained by the availability of data spanning the range of scales necessary to address many of the fundamental questions.

The systematics of population distribution are fundamental components of the concept of human habitat. The spatial distribution of human population in turn influences the interaction with the environment that defines the habitat. The premise is that this spatial distribution is not a random occurrence but is influenced, to some extent, by both physical and nonphysical (e.g., socioeconomic) forces. Geophysical forces are not the only, or perhaps even the most dominant, forces in most cases, but in the context of a global analysis it is necessary to limit the number of dimensions – at least initially. Determining the extent and influence of these factors by quantifying the spatial distribution of human population with respect to a limited number of parameters can help to constrain and test models that seek to explain systematics of human settlement.

Until recently, a systematic global analysis of population distribution would have been severely compromised by the availability of data. Recognition of the importance of scale and dimension coupled with recent technological advances in connectivity, computational capacity and data collection have made it possible to consider fundamental questions about the Earth system in a global context. Specifically, the recent availability of a variety of spatially coded geophysical and demographic data at a wide range of scales facilitates cross disciplinary analyses and hypothesis testing (e.g., Deichmann et al. 2001). In addition, the current renaissance in the field of spatial analysis associated with growing interest in remote sensing and geographic information system (GIS) technology provides the tools and methodologies necessary to capitalize on past investments in monodisciplinary research in the physical and social sciences.

Fundamental relationships between humans and their physical environment have long been assumed without systematic verification while others have only been speculated upon. By quantifying population distributions in a number of dimensions across a wide range of spatial scales it may be possible to establish the existence of systematic patterns which can be used to test and refine competing theories of human population dynamics. We know from observation that certain patterns do exist, but without systematic global analyses there is no way to know if the patterns are related to fundamental processes or just coincidence. If multiscale patterns do exist, and can be explained consistently, then it may be possible to anticipate some consequences of otherwise complex human activities. Quantifying the relationship between human population and the environment could eventually prove valuable to policy-makers who require accurate information about human interaction with the environment. Quantitative analysis may allow spatiotemporal models to be improved such that they might provide sufficient predictive power for scenario testing and projection of consequences of policy decisions. These types of analysis could also incorporate information on the spatial distribution of biodiversity. The works of Bartlett et al. (2000), Cincotta et al. (2000), and Dobson et al. (1997) represent pioneering contributions to our understanding of the spatial relationships between human population and biodiversity. In addition to understanding the relationship between human population and physical environmental factors, the analyses presented here could also be applied to species distributions and proxies for biodiversity to quantify spatial coincidence of biodiversity hotspots and specific environmental conditions.

3.2 Theories of Population Distribution

It is widely acknowledged that population distribution is influenced by political, cultural, socioeconomic, demographic, and geophysical factors. The extent of each influence is variable and is the subject of some controversy as complex modern societies are clearly influenced by other factors also. The interplay between demographic, geophysical, socioeconomic, political, and cultural factors precludes simple explanations of population distribution in most areas. However, one of the fundamental tenets of complex systems theory is the notion of universality and the existence of consistent patterns observed in a variety of individual circumstances (Haken 1978). Some anthropological theories of early human habitat expansion are based almost entirely on environmental and demographic factors (see Hassan 1981 for a review). Modern theories of land use (e.g., Turner et al. 1990) and urban growth (e.g., Batty and Longley 1994) incorporate additional socioeconomic dimensions but are generally limited to local and occasionally regional extent.

Most anthropologic and archeological theories of early human habitat expansion are, by necessity, qualitative. Some exceptions are given by Hassan (1981). Extensive treatments of the general characteristics of the human habitat discuss the role of agriculture (e.g., Cipolla 1970; Reader 1988) and technology (Headrick 1990) on population distribution. Current understanding of population growth and expansion recognizes the importance of agricultural and technological transitions (e.g., Kates et al. 1990; Whitmore et al. 1990) and cross cultural exchange (Diamond 1997). Demographic inquiries of population growth, while quantitative, have been limited to either highly localized estimates (e.g., Wrigley and Schofield 1981) or global cumulative estimates (e.g., Demeny 1990; Cohen 1995); however, in both types of inquiry the spatial distribution of population growth is generally not considered. Recent analyses of urban population distribution (discussed below) provide quantitative descriptions of population distribution but tend to focus on economic and behavioral factors influencing local or regional population distribution.

A variety of mathematical models have been developed to describe allegedly universal characteristics of population distribution at different scales. Early models focused on settlement size distributions in a nonspatial context (e.g., Zipf 1949). Subsequent allometric models incorporated space in a Euclidian sense (e.g., Clark 1951; Nordbeck 1971). Current approaches to urban modeling recognize the non-Euclidian character of urban population distribution and growth (e.g., Batty and Longley 1994; Zanette and Manrubia 1997; Makse et al. 1998). These models focus on the functional analytic form (usually exponential) but include proportionality constants that allow them to be extended to incorporate other modulating factors. Not all of these models are mutually exclusive and most share the property of exponential proportionality. Diffusion-limited aggregation (e.g., Batty and Longley 1994) and correlated percolation (e.g., Makse et al. 1998) account for non-Euclidian geometries with noninteger exponents. In each case, the model was designed to describe some scaling property of population distribution. Many models describe the form of the population distribution independent of geophysical, demographic, and socioeconomic forces. Spatial models are usually tested by comparison to individual cities. To my knowledge, only rank-size models (e.g., Zipf 1949) have been tested at global scales (Berry 1971).

In the context of ecology and biodiversity, it is logical to consider human population with respect to the physical characteristics of the environment that presumably influence the distributions of other species and habitats. Systematic patterns in population distribution suggest order, and perhaps predictability. In an analysis of census data, Cohen (1997) has examined population density distributions and found similar patterns at global, national, and state levels of aggregation for the US and suggested that the distribution of human population density by area may be self-similar (scale independent) over a 1,000-fold range of areas. This conjecture was based on a similarity in the Lorenz curves of areal distribution of population density and land area for the world, the United States, and New York State. If population density does scale in some predictable self-similar manner then this should have direct implications for the impact of human populations on their immediate environment and may allow for assessments of potential political and socioeconomic consequences of migration and population growth to be made. Once global population distribution has been quantified across a wide range of spatial scales, it may be possible to examine the applicability of population distribution models at global, regional, and local scales and to test the sensitivity of these models to basic geophysical factors across a range of socioeconomic, political, and cultural settings. At present, available census data are not sufficiently uniform or detailed to conduct such an inquiry. In the meantime, however, an investigation of the fundamental aspects of global population distribution can illustrate the relationship of human population distribution to some of the basic environmental factors that may influence it as well as the distributions of other species.

3.3 Spatial Distribution of Modern Human Population

3.3.1 Census Data

For most applications, the spatial distribution of human population must be inferred from proxies. Gridded compilations of census data provide spatially explicit numerical estimates of human inhabitants, but the accuracy and spatial resolution vary considerably from one country to another. Results presented here are derived from the Gridded Population of the World (GPW2) dataset (Center for International Earth Science Information Network et al. 2000). At the time this analysis was conducted (2003), GPW2 provided the most spatially detailed representation of population data available. The current generation of the GPW (version 3) is based on considerably higher spatial resolution data for many parts of the world but the overall distribution responsible for the patterns discussed here is consistent with GPW2. In the context of this discussion, population distribution products like Landscan (http://www.ornl.gov/sci/landscan/) would be considered population models because they rely on numerous assumptions that disaggregate population relative to the census data from which they are partially derived. In contrast, the GPW model assumes only that population is uniformly distributed within their respective administrative units (the minimum astonishment assumption) and uses no other information to assign location to population. Gridded estimates of population count and population density (people km^{-2}) are based on a compilation of populations and areas of 127,105 political or administrative units derived from censuses and surveys (see Fig. 3.1). The range of census years used was 1967–1998, and all populations were projected to a common base year of 1990 to yield a global population of 5,204,048,442 people. The uncertain accuracy of the census data implies considerable spurious precision in this total. Population counts are assumed to be distributed uniformly within each administrative unit and sampled on a grid of 2.5 arcmin (2.5′) quadrilaterals (Deichmann et al. 2001). The median spatial resolution of the input census data is 31 km. Detailed description of the spatial properties of this dataset is provided by Deichmann et al. (2001) and Small and Cohen (1999, 2004). For the purposes of the present analysis, all populated areas included in census enumerations are considered potentially habitable land area. This excludes Antarctica and some areas at high northern latitudes.

Human population is strongly clustered in space. Estimates derived from GPW2 indicate that population density varies by more than six orders of magnitude (Small and Cohen 1999, 2004). This is a minimum estimate, because populations are assumed to be uniformly distributed within administrative units. Night light data (discussed below) and experience suggest that there is considerable clustering at finer spatial scales. Population maps are presented on a logarithmic scale to better represent the wide range of population densities. The maps in Figs. 3.1 and 3.2 represent the distribution of large, sparsely populated regions well but many of the small, high density areas are difficult to resolve at the global scale. For this reason, it is difficult to appreciate the extent of spatial clustering represented in the dataset. The clustering

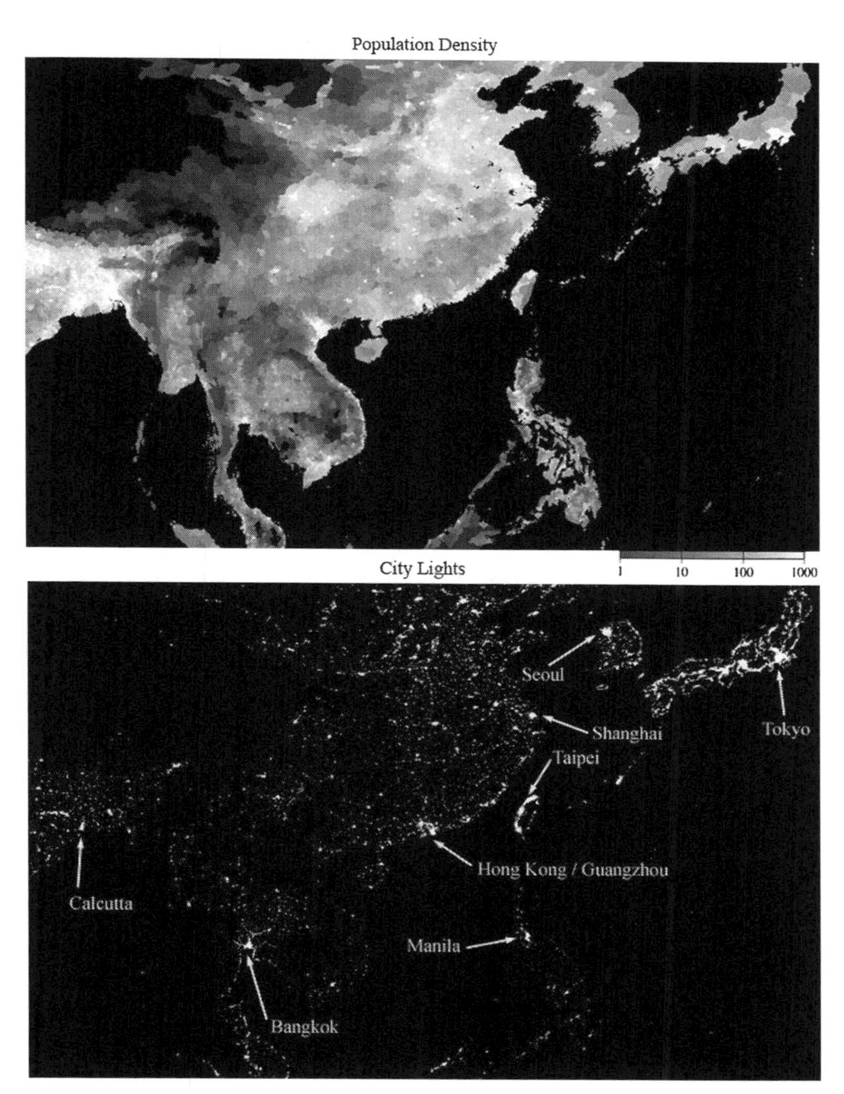

Fig. 3.2 Population density and city lights for southeastern Asia. Note the disparity between extensive dense populations with relatively little lighted area on deltas in comparison to extensively developed coastal zones (Small 2004)

implies that at global scales the distribution of human population is spatially correlated. The degree of autocorrelation is likely scale-dependent, but this is difficult to quantify accurately because the resolution and accuracy of the population data vary spatially. This issue is discussed in greater detail in Small and Cohen (2004).

The extent of population clustering can be summarized with a Lorenz curve for the spatial distribution of population in the GPW2 dataset (Fig. 3.3). This curve

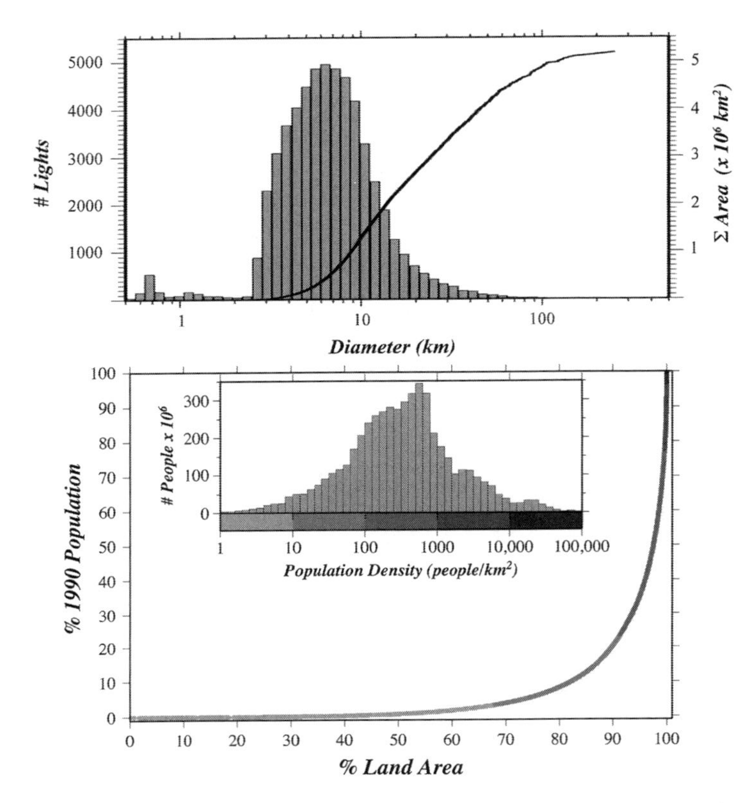

Fig. 3.3 Global distribution of people and settlements as functions of population density and land area. *Top figure* shows that almost all lighted settlements are <20 km diameter but those that are larger account for more than half the total lighted area. *Bottom figure* quantifies clustering of population. *Inset histogram* indicates that most people live at densities between 100 and 1,000 people km^{-2} but the density-shaded spatial Lorenz curve shows while the densest 50% of population occupy less than 3% of land area at densities greater than 300 people km^{-1} (Small 2004)

shows the cumulative population as a function of the cumulative land area that it occupies when the quadrangles are ranked from lowest to highest population density. It is referred to as a spatial localization function to distinguish it from Lorenz curves describing population distribution with respect to nonspatial parameters. The curvature of the function indicates the degree of spatial localization. A linear localization function depicts a uniform distribution over all available area. It is apparent from the curvature of the localization function that human populations are strongly clustered at the global scale. Figure 3.3 indicates that at least 50% of the 1990 human population occupied less than 3% of Earth's potentially habitable land area. The converse is not necessarily true, however. Because there are often local settlements within the low density administrative units, the sparsely populated areas have low densities on average but may contain small concentrations of higher density settlement that are not spatially represented in the census data. Nonetheless, it is still accurate to say that half of Earth's potentially habitable land area is sparsely populated, with less than 2% of the human population at densities of less than 10 people km^{-2}.

3.3.2 *Night Lights*

Satellite-derived maps of temporally stable night lights provide an additional source of information on the spatial extent of human development. The Defense Meteorological Satellite Program/Operational Linescan System (DMSP/OLS) has provided observations of the location and frequency of occurrence of Visible and Near Infrared (VNIR) emissions from fires, lighted human settlements and other anthropogenic sources since 1972 (Croft 1978). These data have been compiled and processed to provide a unique measure of the location and spatial extent of lighted human settlements by Elvidge et al. (1997a, 1999). Temporal persistence distinguishes stable lights associated with settlements from intermittent emissions from fires. The global distribution of lighted settlements is shown in Figs. 3.2 and 3.3. From this dataset it is possible to derive estimates of the locations (centroids) and areas of lighted settlements worldwide. In discussions of distance and resolution, it is more intuitive to think in terms of a linear dimension rather than of area (A). Thus, the size of lighted settlements is given in units of equivalent circular diameter (d_c), defined here as

$$d_c = 2(A/\pi)^{1/2}. \tag{3.1}$$

The size distribution and cumulative area of the light dataset is shown in Fig. 3.3.

The use of lighted areas as a proxy for human settlement is subject to a number of caveats. Detailed comparison of the stable light dataset to higher resolution (30 m) Landsat imagery indicates a detection threshold below which increasing numbers of smaller lighted settlements are not imaged (Small et al. 2005). The limited spatial resolution (2.5 km) and sensitivity of the OLS sensor precludes detection of many small light sources (Elvidge et al. 2004). Atmospheric scattering and spatial uncertainty in geo-positioning also diminish the spatial accuracy of satellite-detected night lights. Because the spatial extent of individual lighted areas is known to be somewhat greater than the actual built up area of many settlements (Elvidge et al. 1997a), discussions of area and dimension must account for overestimation. Recent comparisons of lighted areas with Landsat imagery (Elvidge et al. 2004; Small et al. 2005) indicate that lighted area linearly overestimates built-up area by approximately a factor of two. In units of equivalent diameter this translates to an overestimation by a factor of 1.3. However, more recent analyses reveal that the spatial overextent of lighted area generally corresponds to areas of lower intensity of development (Small et al. 2010).

In spite of these caveats, stable lights do provide a valuable complement to census-derived proxies for human settlement patterns. The median area of the 54,478 contiguous lighted settlements in this dataset is 33 km^2, corresponding to a median settlement diameter of 6.5 km (circular equivalent). In comparison to the population weighted median area of 961 km^2 and equivalent resolution of 31 km for the census tracts used in GPW2, the night light dataset offers considerably higher spatial resolution and, therefore, provides complementary estimates of the number and size of more developed human settlements. Lighted area is moderately correlated ($r^2 = 0.68$) with population (Sutton et al. 1997, 2001). Because of the considerable variability in the relationship between lighted area and population, I draw no conclusions about the number of people living within the lighted areas. Rather, the size and location of the lighted settlements are used as spatially explicit indicators of local concentrations of population that are associated with significant economic investment in infrastructure.

The area and size distribution of light sources provides a measure of spatial clustering of population. The size distribution of the lighted areas indicates the importance of large conurbations relative to the more numerous smaller settlements. Less than 5% of the 54,478 light polygons (circumscribing the stable lighted areas) have circular equivalent diameters larger than 20 km, yet they account for 50% of the lighted area (see Fig. 3.3). The total lighted area (5.18×10^6 km^2) accounts for less than 4% of the 130,582,040 km^2 of inhabited land area. Adjusting for the spatial overestimation of lighted area suggests that the settlements imaged occupy less than 2% of inhabited land area. Accounting for undetected smaller settlements would increase this area but without knowing the size frequency distribution of undetected settlements it is not possible to make a meaningful estimate. If lighted area is used as a proxy for urban area the satellite data indicate that at least 2.5 million km^2 are occupied by urban and developed areas (including airports, oil production facilities, and other sparsely populated lighted areas). This is certainly a significant underestimate, because of the reasons noted above and because many small settlements are not illuminated at night. The size frequency distribution of lighted settlements does, however, provide a reasonably accurate description of the scaling properties of larger human settlements. In the context of this analysis, the combination of lighted settlements and moderate resolution census data can distinguish between densely populated rural areas and more heavily developed centers of economic activity.

3.4 Continental Physiography

In the context of this discussion, continental physiography refers to the basic physical properties of the landscape. Here we consider elevation above sea level and proximity to permanent rivers and seacoasts. Landscape is also characterized by morphologic properties such as slope and aspect. These properties are strongly scale-dependent. A scale-based analysis of higher order morphologic properties is beyond the scope of this study. In many areas, the spatial resolution of the census data is not sufficient to support a scale-based analysis. Nonetheless, these properties may be as important as the basic properties considered here. The population and land area relationships summarized here are derived from the more detailed analysis given in Small and Cohen (2004).

Coastal proximity is calculated as horizontal distance to the nearest coastline at each point for which a population estimate is available. The coastline used in the analysis is the Global Self-consistent Hierarchical High Resolution Shoreline digital coastline file (Wessel and Smith 1996), which consists of 10,390,243 points worldwide. This product was assembled from the World Vector Shoreline and the US Central Intelligence Agency World Data Bank II datasets. The largest uncertainty with the coastline data is the definition of the coastline up rivers/estuaries. Elevation is calculated as the vertical distance to mean sea level at each point for which a population estimate was available. Elevation estimates are derived from global, 30 arcsec (30″) gridded elevations provided by the USGS EROS Data Center (http://edcwww.cr.usgs.gov/landdaac/landdaac.html). The 30″ elevation model was derived from Defense Mapping Agency (DMA) map products and DMA digital terrain elevation data (DTED). The vertical uncertainty in the original 30″ elevation model is generally in excess of one vertical meter but varies within the dataset as a result of the diversity of sources of the elevation data (Danko 1992). Fluvial proximity was calculated as distance from the nearest permanent river, as defined by the Digital Chart of the World (US Government 1993). This dataset has limited ability to resolve small tributaries and should not be interpreted as representative of all flowing water sources. The accuracy and resolution of the coastline, elevation, and population data impose important constraints on the conclusions that can be drawn from these global datasets. Calculating each of these parameters (elevation, coastal proximity, fluvial proximity) for each gridded population estimate results in distributions of population and land area as functions of each parameter. A detailed discussion of this analysis and its results is given in Small and Cohen (1999, 2004). Maps of the physiographic parameters are shown in Fig. 1 and available online at: http://www.LDEO.columbia.edu/~small/population.html.

Human population is strongly localized with respect to the physiographic parameters considered here. Figure 3.4 indicates that both population and lighted settlements occur in greatest abundance at low elevations in close proximity to rivers and sea coasts and that all diminish rapidly with distance. In part, this is a consequence of the distribution of available land area (Fig. 3.4). Nonetheless, when population at a given distance (or elevation) is normalized by the available land area the resulting average population densities still indicate pronounced clustering within 100 km of rivers and sea coasts and within 100 m of sea level (Fig. 3.4). This localization is even more pronounced for lighted settlements as indicated by significant increases in lighted area from large sources at the smallest elevations and proximities. Average population densities have maxima corresponding to the same peaks observed in the distributions of population and lighted area but also highlight other peaks not apparent in the other distributions. The peak in average density at 2,300 m elevation corresponds to the densely populated Mexican plateau.

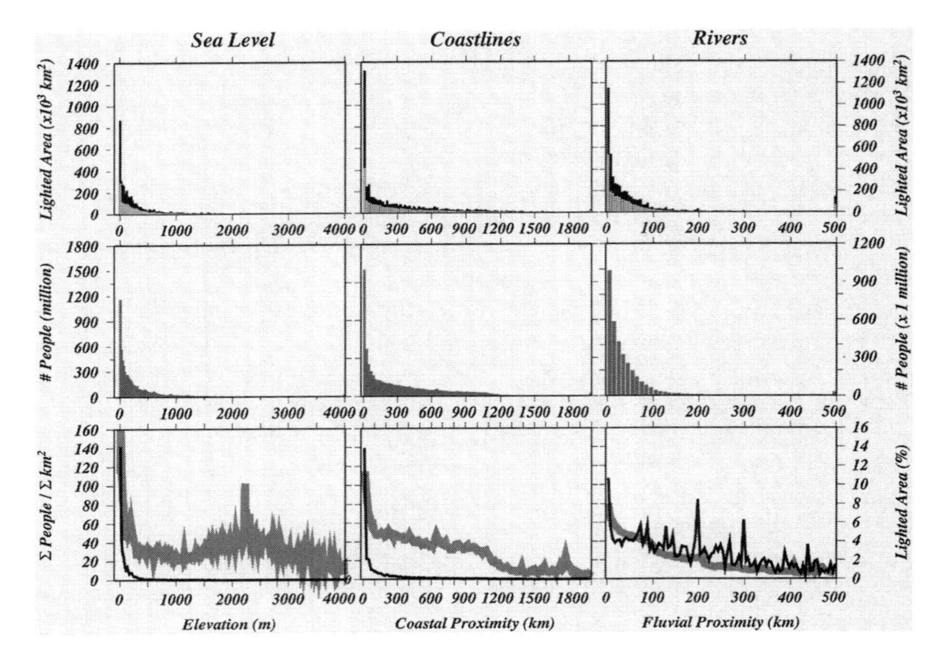

Fig. 3.4 Distributions of population and lighted area relative to continental physiographic features. Lighted area (*top row*) is strongly clustered with the both total lighted area (*darker*) and lighted settlements smaller than 20 km diameter (*lighter*) diminishing rapidly with elevation and distance from rivers and coastlines. The proportion of smaller settlements increases with distance and elevation. Total population (*second row*) shows a similar clustering. In part, this is a result of the distribution of available land area at each distance (or elevation). When population at each distance (or elevation) is normalized by the available area the resulting average population densities (*bottom row gray*) are much higher within 100 km of coastlines and within 100 m of sea level. The decrease with distance from rivers is more gradual. Percent area lighted (*bottom row black*) is even more strongly localized suggesting that the density peaks are associated with urban areas (Small 2004). Details of the population and land area analysis are given in Small and Cohen (2004)

Although the density approaches the high values occurring at the lowest elevations, this peak represents far fewer people (518 million vs. 4.8 million).

3.5 Climate

Climatic factors are often assumed to influence human habitation patterns at continental scales. While climatic extremes can obviously preclude human habitation, the importance of "favorable" climatic conditions to human population distribution has been the subject of some debate (e.g., Huntington 1927). This analysis makes no such assertions but merely attempts to quantify the relationship and discuss whatever consistencies may emerge. The analysis considers basic characteristics of temperature and precipitation in the form of annual averages and annual ranges of each. Because climate and weather change constantly, both annual and interannual variability are also considered. The regularity of annual cycles is relatively deterministic. Interannual variability is inherently more stochastic and, therefore, less predictable in most environments. For this reason, we compare the geographic distribution of both annual ranges and interannual anomalies for temperature and precipitation.

As with landscape, climate is also manifest by scale-dependent properties. Climatic variables have the added complexity of temporal variability across a wide range of scales. Many of these properties can be represented in terms of the phase and the amplitude of temperature and precipitation cycles. A thorough analysis of these higher order climate parameters is beyond the scope of this study, but these parameters may be as important as those considered here. The multidimensional analysis described by Small and Cohen (1999, 2004) could be extended to include these dimensions as well as the higher order physiographic parameters discussed previously. The population and land area relationships summarized here are derived from the more detailed analysis given in Small and Cohen (2004).

Climatic parameters discussed in this study were derived from global climatologies compiled by Mitchell and Jones (2005). Gridded monthly averages of 12,092 temperature stations and 19,295 precipitation stations, compiled over the years 1960–2002 were used to calculate annual average and annual range (maximum minus minimum) for each 0.5° grid cell provided by Mitchell and Jones (2005). The annual averages and ranges discriminate the primary climatic divisions observed at global scales. These climatic data do not resolve distinct microclimates that may exist at scales finer than 0.5° (about 55 km at the equator). Other important climatic variables (e.g., wind, frost days, cloud cover, potential evapotranspiration) are not resolved by these data. A detailed discussion of this analysis and its results is given in Small and Cohen (1999, 2004), and color maps of the climatic parameters are shown in Fig. 3.5 (available online at: http://www.LDEO.columbia.edu/~small/population.html).

Interannual variability is quantified as anomalies relative to the expected temperature or precipitation at the monthly or annual time scales. Because interannual variability generally scales with the magnitude of the annual mean or range, we also consider normalized interannual anomalies as a measure of the stochastic departure from expected climate. Mean anomalies are calculated as the 42-year mean of the difference between each individual monthly average and the 42-year monthly average climatology. To derive normalized anomalies the mean temperature anomaly is divided by the annual temperature range and the monthly mean precipitation anomaly is divided by the 42-year average precipitation for each month.

The distributions of population and lighted settlements are not strongly localized with respect to any of the climatic parameters considered here. While all of the distributions have peaks (Fig. 3.6), none is as strongly skewed as any of the physiographic parameters shown in Fig. 3.4. Even when land area distributions are considered, the average population densities are not as strongly localized as those for physiographic parameters. The distributions of lighted area and population differ markedly for average temperature and annual variability of precipitation. These differences of distribution highlight distinctions between urban and rural

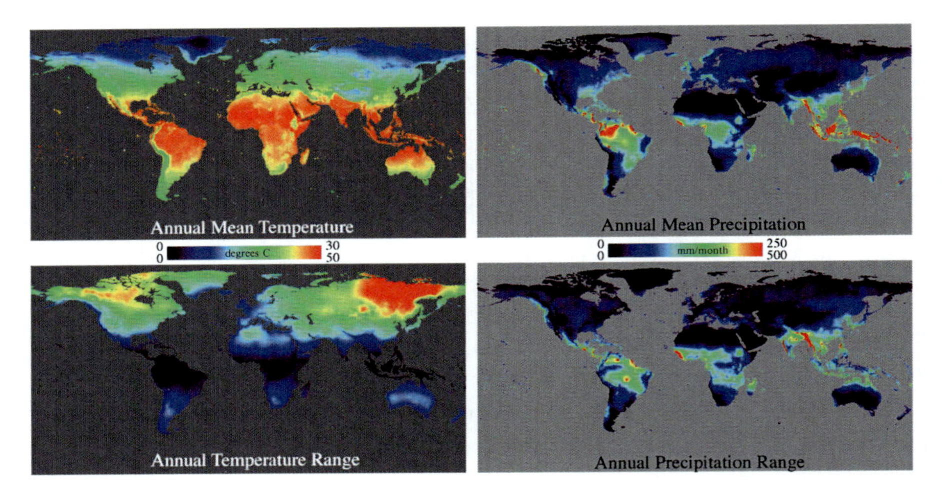

Fig. 3.5 Annual temperature and precipitation. The annual means and annual ranges of monthly average temperature and precipitation from 1960 to 2002 are drawn from Mitchell and Jones (2005). These spatial patterns represent the relatively stable aspects of the climate system on a scale of a human lifetime. The ranges represent the relatively predictable annual component of climatic variability

population and development. The large peak in population at annual average temperatures of 25°–26° is not reflected in a comparable peak in lighted area. Similarly, the broad secondary peak in population at annual precipitation ranges of 2,000 mm per year is not accompanied by a similar peak in lighted area. The causes and implications of these differences are discussed below (Fig. 3.7).

3.6 Implications

Complex spatial distributions of population and urban settlements produce consistent patterns in geophysical parameter spaces used to quantify human habitat (Small 2004). The population densities shown in Figs. 3.1 and 3.2 highlight spatial clustering at a range of scales. The spatial distribution of the stable lights in these figures shows further clustering at finer spatial scales. The complex spatial distributions of population are related, in part, to the spatial complexity of the distribution of land area with geophysical conditions conducive to human habitation. The simple analyses presented here demonstrate how projections of multidimensional distributions in geophysical parameter space can reveal consistencies that are not obvious in geographic space.

The disparities between the climatic and physiographic distributions of population and lighted urban areas reflect fundamental characteristics of the modern human habitat. Adaptation to climate has allowed extensive settlement of all nonpolar climatic zones and limited settlement of more extreme polar and desert environments, yet the global distribution of population is strongly localized with respect to physiographic aspects of the landscape. Figure 3.6 indicates that human

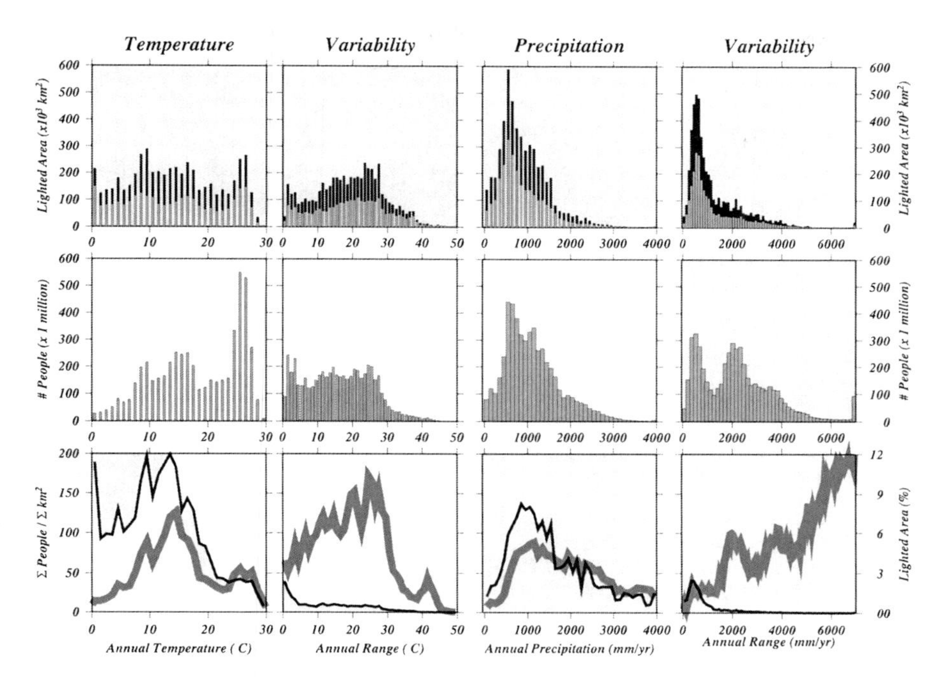

Fig. 3.6 Distributions of population and lighted area relative to climatic variables. Lighted area (*top row*) and population (*second row*) have similar distributions except at 26° annual average temperature and 2,000 mm per year annual precipitation range where large rural populations occur with relatively little urban development. As with physiographic features, this is partly a result of the distribution of available land area. Normalizing population and lighted area by available land area shows how average population or urban density varies with climate. Average population densities (*bottom row*, *gray*) are much more dispersed than those of the physiographic distributions shown in Fig. 3.4. The disparity between the average densities and percentage lighted area (*bottom row*, *black*) emphasizes the difference between urban and rural populations and highlights the importance of tropical monsoons to large rural populations dependent on subsistence agriculture. Details of the population and land area analysis are given in Small and Cohen (2004)

population is distributed over a wide variety of climatic conditions with little evident clustering. However, the spatial scales over which climate varies are often considerably larger than the physiographic proximities (<100 km) indicated by Fig. 3.5. This indicates that while humans have adapted to a wide range of climates, the majority of the population is clustered with respect to physiographic landscape. The implication is that landscape physiography imposes a stronger constraint than climate on human habitation.

While the global distribution of deterministic climatic patterns has been studied extensively by geographers and climatologists in terms of climatic classifications (see Thornthwaite 1948 for a review), the global distribution of stochastic variability has received less attention. The regions with the largest mean anomalies tend to be sparsely populated but many of the areas with the largest normalized anomalies are densely populated. Absolute anomalies in temperature tend to be at high

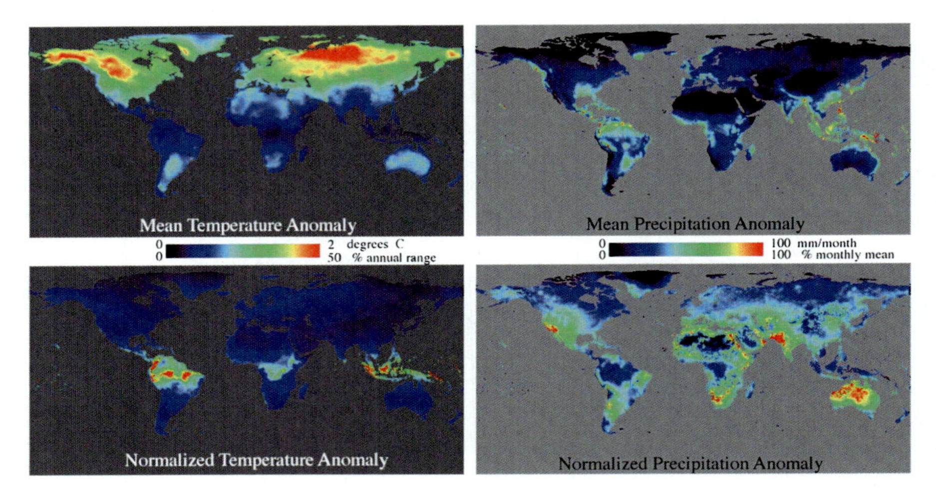

Fig. 3.7 Interannual variability of temperature and precipitation. Average monthly temperature and precipitation anomalies (*top*) from 1960 to 2002 from Mitchell and Jones (2005) depict the stochastic component of climate. Mean monthly anomalies are correlated with seasonal temperature range and total precipitation. Climatic variability may have greater impact when the unpredictable component is large compared to the predictable annual cycle. When normalized by the magnitude of the annual temperature range or mean monthly precipitation, the locations and magnitudes of the anomalies change considerably. These spatial patterns represent the less predictable aspects of the climate system on a scale of a human lifetime

northern latitudes but large normalized anomalies occur along the west coast of South America, central and eastern Africa, Sri Lanka, and the Austral-Asian Archipelago. Absolute anomalies in precipitation are large throughout the tropics and subtropics but the largest normalized interannual anomalies occur at the peripheries of the arid and semiarid subtropics affected by the annual movement of the Intertropical Convergence Zone (ITCZ). Many of these areas are densely populated by rural communities dependent on subsistence agriculture.

With regard to climate change, we consider the implications of the geographic distribution of climatic anomalies occurring over the past 40 years. Precipitation anomalies, in particular, have implications for human health and socioeconomic stability in many areas. In less developed areas where populations depend on subsistence agriculture variability in precipitation has well-known implications such as drought and crop failure. In more developed areas like eastern Australia and the southwestern United States droughts affect nonagricultural areas through wild fire. In terms of future climate change, the high interannual variability throughout the most densely populated parts of Africa has grim implications if this variability remains high or increases. Similar conclusions could be drawn for the region between the Persian Gulf and India northward through Central Asia. While climatic variability, particularly in precipitation, can have a strong influence on environmental stresses, socioeconomic, cultural, and political factors must also be considered in such scenarios.

The combination of climatic and landscape characteristics conveys a variety of environmental benefits. Analysis of bivariate physiographic distributions highlights the dense populations associated with deltas and depositional river basins (Small and Cohen 1999, 2004). The Indus, Ganges–Brahmaputra, Irrawaddy, Mekong, Red, Pearl, Huang, Huai, and Hai river basins and Szechuan basin provide graphic examples. Similar analyses of coastal zones reveal clustering of lighted settlements within 10 km of seacoasts in contrast to high rural population densities extending inland at low elevations (Small and Nicholls 2003). Coastal settlements adjacent to the Mediterranean and South China seas provide examples of this clustering. Populations have also been shown to cluster in proximity to volcanic zones (Small and Naumann 2001), as illustrated in Central America; East Africa; and the Indonesian, Japanese, and Philippine archipelagos. Clustering in volcanic zones generally coincides with a combination of physiographic, geologic, and micro-climatic conditions favorable to agriculture. The environmental benefits of river valleys, deltas, and coastal zones are well known and require no further explanation.

The spatial disparity between urban and rural settlement patterns is also strongly localized with respect to physiography, but is generally consistent with overall population distributions for climatic parameters. The proximity of lighted areas to rivers and sea coasts is more pronounced than the proximity of population overall (Fig. 3.5), while the distributions are similar with respect to elevation. This reflects intense economic development near trade routes (Smith 1776; Gallup et al. 1999) in contrast to rural agricultural development of low elevation river basins. Conversely, distributions of population and lighted area are generally similar for the climatic parameters shown in Fig. 3.6. Notable exceptions are again related to dense rural populations in agricultural areas without large lighted areas. The prominent peak of population centered at 26°C average temperature is largely a result of the dense rural populations of the river plain of the Ganges. The broad peak at 2,000 mm per year annual precipitation range is related to the dense rural populations of the sub-Saharan Sahel. The prominent shoulder on this peak extending to higher annual ranges reflects the dense populations in Southeast Asian monsoon regions.

Dense populations not accompanied by lighted areas generally correspond to rural areas of intensive agriculture while lighted areas are associated with centers of economic development. At a national level, lighted area is correlated with GDP and energy consumption (Elvidge et al. 1997b). The analysis presented here and by Small (2004) highlights the difference between urban and rural populations. This difference is significant in light of United Nations projections that most of the population growth in the next 50 years will occur in moderate-sized cities in developing countries (United Nations Population Division 2002). As many of these developing countries are in the tropics, population growth and urbanization are likely to occur in tropical biomes associated with much of the world's biodiversity.

The spatial coincidence of biodiversity hotspots and human population has different implications in different cultural, political, and socioeconomic settings. In developed countries, there are often greater resources available for protection of biodiversity hotspots, but the development pressures may focus specifically on physical environments where the hotspots are concentrated. A detailed analysis by Bartlett et al. (2000) found that nontraditional settlements in the United States tend to be associated with desert and coastal ecosystems coincident with higher concentrations of threatened and endangered species. The work of Dobson et al. (1997) found that centers of endemism tend to be very limited in number and area and rarely overlap with each other but often overlap with centers of anthropogenic activity in the United States. The concentration of large proportions of endangered species from different taxa in relatively small areas favors conservation in developed countries where resources are available to protect these areas. However, global conservation success may well rest on efforts to maintain biological diversity in developing countries, where the task can be more challenging.

If future population growth occurs primarily in urban areas of developing countries, the result may be selective pressure on specific habitats and biomes. The modern proximity of lighted settlements to rivers and sea coasts has direct implications for riparian habitats, wetlands, and coastal zones. These implications may be positive or negative, depending on how the growth is managed. If the UN projections are correct, the moderate-sized cities of today could be the megacities of tomorrow. This would imply further spatial clustering of human population but would not necessarily imply diminished impact on surrounding ecosystems. Urban populations consume natural resources that must be extracted from surrounding areas as well as global networks (Small et al. 2010). If populations migrate to urban areas but depopulated rural areas continue to be used for agricultural or industrial production, the impact on biodiversity could be equal to or greater than the impact from dispersed rural communities. While indigenous local populations may have cultural or socioeconomic incentives to conserve habitats in biodiversity hotspots, the same may not be true in the drivers of large-scale agricultural or industrial land use in developing countries. Multidimensional spatial analyses like those presented here could be used to inform broader cross-disciplinary analysis of future population growth and development scenarios. Anticipating the potential consequences of accelerated human habitation of sensitive biomes may help minimize negative impacts and unintended consequences of unmanaged growth.

Acknowledgments Much of the research presented in this chapter was supported by the University Consortium for Atmospheric Research (UCAR) Visiting Scientist Program and by NASA Socio Economic Data and Applications Center (SEDAC). The author also gratefully acknowledges the support of the Palisades Geophysical Institute and the Doherty Foundation.

References

Bartlett JG, Mageean DM, O'Connor RJ (2000) Residential expansion as a continental threat to U.S. coastal ecosystems. Popul Environ 21:429–468

Batty M, Longley P (1994) Fractal cities. Academic, San Diego

Berry B (1971) City size and economic development. In: Jakobson L, Prakash V (eds) Urbanization and national development. Sage, Beverly Hills, CA

Boserup E (1965) The conditions of agricultural growth. Aldine, Chicago

Boserup E (1981) Population and technological change: a study of long term trends. University of Chicago Press, Chicago

Center for International Earth Science Information Network (CIESIN), Columbia University; International Food Policy Research Institute (IFPRI), and World Resources Institute (WRI) (2000) Gridded population of the world (GPW), Version 2. CIESIN, Palisades, NY. http://sedac.ciesin.org/plue/gpw

Cincotta RP, Wisnewski J, Engelman R (2000) Human population in the biodiversity hotspots. Nature 404:990–992

Cipolla CM (1970) The economic history of world population. Penguin, Harmondsworth, UK

Clark C (1951) Urban population densities. J R Stat Soc 114:490–496

Cohen JE (1995) How many people can the earth support? WW Norton, New York

Cohen JE (1997) Conservation and human population growth: what are the linkages? In: Pickett STA, Oldfeld RS, Shachak M, Likens GE (eds) The ecological basis of conservation. Chapman and Hall, New York

Cohen JE, Small C (1998) Hypsographic demography: the global distribution of population with altitude. Proc Natl Acad Sci 95:14009–14014

Croft TA (1978) Nighttime images of the earth from space. Sci Am 239:86

Danko DM (1992) Defense Mapping Agency product specifications for digital terrain elevation data (DTED), 2nd edn. GeoInfo Sys 2:29

Deichmann U, Balk D, Yetman G (2001) Transforming population data for interdisciplinary usages: from census to grid. CIESIN, Palisades, NY

Demeny P (1990) Population. In: Turner BLI, Clark WC, Kates RW, Richards JF, Mathews JT, Meyer WB (eds) The earth as transformed by human action. Cambridge University Press, Cambridge, UK

Diamond J (1997) Guns, germs, and steel: the fates of human societies. Norton, New York

Dobson AP, Rodriguez JP, Roberts WM, Wilcove DS (1997) Geographic distribution of endangered species in the United States. Science 275(5299):550–553

Elvidge CD, Baugh KE, Kihn EA, Kroehl HW, Davis ER (1997a) Mapping city lights with nighttime data from the DMSP operational linescan system. Photogramm Eng Remote Sens 63(6):727–734

Elvidge CD, Baugh KE, Kihn EA, Kroehl HW, Davis ER, Davis CW (1997b) Relation between satellite observed visible-near infrared emissions, population, economic activity and electric power consumption. Int J Remote Sens 18(6):1373–1379

Elvidge CD, Baugh KE, Dietz JB, Bland T, Sutton PC, Kroehl HW (1999) Radiance calibration of DMSP-OLS low-light imaging data of human settlements. Remote Sens Environ 68(1):77–88

Elvidge CD, Safran J, Nelson IL, Tuttle BT, Hobson VR, Baugh KE, Dietz JB, Erwin EH (2004) Area and position accuracy of DMSP nighttime lights data. In: Lunetta RS, Lyon JG (eds) Remote sensing and GIS accuracy assessment. CRC, New York

Gallup JL, Sachs JD, Mellinger A (1999) Geography and economic development. Int Reg Sci Rev 22(2):179–232

Haken H (1978) Synergetics: an introduction. Springer, Berlin

Hassan FA (1981) Demographic archaeology. Academic, New York

Headrick DR (1990) Technological change. In: Turner BLI, Clark WC, Kates RW, Richards JF, Mathews JT, Meyer WB (eds) The earth as transformed by human action. Cambridge University Press, Cambridge, UK

Huntington E (1927) The human habitat. Van Nostrand, New York

Kates R, Turner BLI II, Clark WC (1990) The great transformation. In: Turner BLI, Clark WC, Kates RW, Richards JF, Mathews JT, Meyer WB (eds) The earth as transformed by human action. Cambridge University Press, Cambridge, UK

Livi-Bacci M (1997) A concise history of world population. Blackwell, Oxford, UK

Makse HA, Andrade JS, Batty M, Havlin S, Stanley HE (1998) Modeling urban growth patterns with correlated percolation. Phys Rev E 58(6):7054–7062

Malthus TR (1798) An essay on the principle of population. Johnson, London

Mitchell TD, Jones PD (2005) An improved method of constructing a database of monthly climate observations and associated high-resolution grids. Int J Climatol 25(6):693–712

Nordbeck S (1971) Urban allometric growth. Geogr Ann B 53:54–67

O'Neill B, Balk D (2001) Projecting World Population Futures. Population Bulletin 56(3), also available at http://www.prb.org/Content/NavigationMenu/PRB/AboutPRB/ Population_Bulletin2/World_Population_Futures.htm

Reader J (1988) Man on Earth. Collins, London

Sachs J (1997) The limits of convergence: nature, nurture and growth. The Economist 343:19–22

Small C (2004) Global population distribution and urban land use in geophysical parameter space. Earth Interact 8:1–18

Small C, Cohen J (2004) Continental physiography, climate and the global distribution of human population. Curr Anthropol 45:269–277

Small C, Cohen JE (1999) Continental physiography, climate and the global distribution of human population, Proceedings of the International Symposium on Digital Earth, Beijing China, pp. 965–971

Small C, Naumann T (2001) Holocene volcanism and the global distribution of human population. Environ Hazards 3:93–109

Small C, Nicholls R (2003) A global analysis of human settlement of coastal zones. J Coastal Res 19(3):584–599

Small C, Pozzi F, Elvidge C (2005) Spatial analysis of global urban extent from DMSP-OLS night lights. Remote Sens Environ 96:277–291

Spatial scaling of stable night lights Christopher Small, Christopher D. Elvidge, Deborah Balk, Mark Montgomery, In Press, Corrected Proof, Available online 16 October 2010

Smith A (1776) An inquiry into the nature and causes of the wealth of nations. Modern Library, New York

Sutton P, Roberts C, Elvidge C, Meij H (1997) A comparison of nighttime satellite imagery and population density for the continental united states. Photogramm Eng Remote Sens 63 (11):1303–1313

Sutton P, Roberts D, Elvidge C, Baugh K (2001) Census from heaven: an estimate of the global human population using night-time satellite imagery. Int J Remote Sens 22(16):3061–3076

Thornthwaite CW (1948) An approach toward a rational classification of climate. Geogr Rev 38(1):55–94

Turner BLI, Clark WC, Kates RW, Richards JF, Mathews JT, Meyer WB (eds) (1990) The earth as transformed by human action. Cambridge University Press, Cambridge, UK

United Nations Population Division (2002) Urbanization prospects, the 2001 revision. ST/ESA/SER.A/166. United Nations, New York

US Government (1993) Digital chart of the world, digital chart of the world database military specification (MIL-D-89009). Defense Printing Service, Philadelphia. http://www.lib.ncsu.edu/stacks/gis/dcw.html

Vitousek PM, Mooney HA, Lubchenco J, Melillo JM (1997) Human domination of Earth's ecosystems. Science 277:494–499

Wessel P, Smith WHF (1996) A global self-consistent, hierarchical, high-resolution shoreline database. J Geophys Res 101(B4):8741–8743

Whitmore TM, Turner BLI, Johnson DL, Kates RW, Gottschang TR (1990) Long-term population change. In: Turner BLI, Clark WC, Kates RW, Richards JF, Mathews JT, Meyer WB (eds) The earth as transformed by human action. Cambridge University Press, Cambridge, UK

Wrigley EA, Schofield RS (1981) The population history of England, 1541–1871. Edward Arnold, London

Zanette DH, Manrubia SC (1997) Role of intermittency in urban development: a model of large scale city formation. Phys Rev Lett 79(3):523–526

Zipf GK (1949) Human behavior and the principle of least effort. Addison-Wesley, Cambridge, UK

Chapter 4
Behavioral Mediators of the Human Population Effect on Global Biodiversity Losses

Jeffrey K. McKee and Erica N. Chambers

4.1 Introduction

Despite our understandings of sound and tested ecological principles over vast time scales, interpretations of occurrences in the natural world during our modern human slice of geological time is fraught with uncertainty. Yet if we combine time depth from the fossil and archeological records with contemporary data of global reach, we can begin to dissect out the most relevant factors that threaten the future of all levels of biodiversity on this planet. It is our contention that the size of the human population and the scale of the human endeavor led to a dramatic rise in extinctions over the past 10,000 years. Continued exponential growth in the human population and our resultant environmental dominance, due to cultural development and ecological contingencies, is rapidly leading to a global mass extinction.

The fossil record of Earth's distant past is instructive, as it is littered with species that have dwindled into extinction. The reasons for past extinctions are many and comprise the topics of rigorous debates among paleontologists and evolutionary biologists. Climatic and environmental changes constantly challenge species of plants, animals, and microbes to find new niches. Novel adaptations to new or altered modes of existence are necessary components of survival. Some groups successfully evolve into new species, involving a "transitional extinction" of the parent species. But more often, the inability to adapt leads to a "terminal extinction", literally a dead end. The complex causes of terminal extinctions are not always easy to discern.

It is not unusual in nature for the rise and success of one species to lead to the downfall of another. Competition can be "red in tooth and claw", as often envisioned. However, it is more common for the effects of a competitor to be profoundly subtle – the product of intricate ecological systems that have developed with evolving components over long periods of time, some on the order of thousands, others millions, of years. The entry of humans or their predecessors into these ecosystems, like that of any other competitor, can thus be expected to have led to a pattern of extinction among certain organisms. Humans were in competition for the finite resources afforded by varied ecosystems – our ancestors' successes in each environment left little for our competitors, and many were vanquished.

R.P. Cincotta and L.J. Gorenflo (eds.), *Human Population: Its Influences on Biological Diversity*, Ecological Studies 214,
DOI 10.1007/978-3-642-16707-2_4, © Springer-Verlag Berlin Heidelberg 2011

Our analysis of the past and present states of global ecological affairs is premised and tested upon the hypothesis that human population density is a major factor in both the losses and threats to other species. Research at the species level of biodiversity can be viewed as a scientific convenience based on widespread availability of data. Research indicates that species are disappearing at a pace possibly 1,000 times that of historic background rates (Pimm et al. 1995). In addition, it should be made clear that also there have been real losses of biodiversity at the genetic level of many species, as surviving populations lessen in numbers. This is an important consideration because the resilience and long-term adaptive capacity of a population is dependent on the genetic variability upon which natural selection can act. Many allelic variations of the species' genes have already gone extinct, even if the species survives. Terminal extinctions become more likely than transitional extinctions, and thus we have already incurred an "extinction debt" for the future (Cowlishaw 1999). As long as our population continues to grow and exert pressure on the natural world, that debt will increase.

A further caveat to species-level studies of biodiversity is that higher levels of biological organization do not automatically get considered. The sustainability of our biosphere also depends on the survival of diverse ecosystems, each of which harbors endemic species as well as key population variants of more widespread species. Yet a study of species, past and present, can still serve as a useful barometer of "biospheric pressure".

4.2 Past Human Population Impacts on Species Biodiversity

The effects of human population growth on species biodiversity may have had a substantial time depth, depending on which of our ancestors one can comfortably call "human". Following the origin and spread of *Homo erectus* circa 1.8 million years ago (mya), there was a substantial decline in mammalian biodiversity in Africa (Behrensmeyer et al. 1997; McKee 2001). From a scientific perspective, it is difficult to attribute these mid-Pleistocene extinctions to *H. erectus*, let alone to the population growth of this species. Yet the coincidence of the increased rate of mammalian extinctions with "human" geographic incursions independently spans across four geographical regions (Klein 2000). Furthermore, our increased body size and the metabolically demanding brain size required greater demand for natural food resources. Thus, the features that allowed our ancestors to compete successfully, and thereby expand their populations, played into the likelihood of a more profound ecological impact on their competitors and prey.

It is reasonable to suggest that by the time our own species, *Homo sapiens*, spread to the new world toward the end of the Pleistocene, human population growth could at least partially account for the overkill of North American megafauna (Alroy 2001). These continental effects of humans on biodiversity took time as human populations slowly reached a critical mass before their impact was great enough to cause extinctions. Islands such as Madagascar, New Zealand, and Hawai'i had elevated levels of species richness combined with smaller habitat sizes such that critical masses were reached more quickly and extinctions followed with greater rapidity (Holdaway and Jacomb 2000; Mlot 1995; Pimm et al. 1995). These global patterns of biodiversity loss led McKee (2003) to attribute many past extinctions to the effects of the growth and spread of human and prehuman populations. Population growth was argued to be a primary *cause*, mediated by aspects of human biology and behavior, as opposed to a spurious correlate of incidental effects.

The impact of human population growth on continental biodiversity accelerated with the origin and spread of agriculture over the past 10,000 years (Redman 1999; McKee 2003), but not without a cost. This lifestyle shift, from nomadic foraging by small bands of people to a group-based sedentary lifestyle, was based on primary food production utilizing monocropping and herding techniques (Armelagos 1990; Barrett et al. 1998). Predictable food supplies altered the birth/death rate equilibrium, resulting in increased population densities (Roberts and Manchester 1997). Although the viability of other species was still impacted by our growing ecological influence, it was mediated in a different way. Rather than directly killing off species through hunting or outcompeting other species for natural food resources, agriculturists promoted wholesale displacement of both plants and animals by utilizing expanses of land for crops and herding. Agricultural lands necessarily became less diverse and less productive in biomass as concentrations of domesticates were grown specifically for human consumption, at the expense of more diverse systems that had evolved to sustain many species.

One of the great bioarcheological ironies is that human health and longevity declined with the origins and spread of agriculture (Larsen 1995). Although reliance on fewer food types decreased nutritional value intake, human populations managed to flourish. Building upon an established base of human "capital", the exponential nature of population growth – even at a slow growth rate – ensured that our numbers increased (McKee 2003). Meanwhile, large mammal extinctions reached an all-time high. For example, in South Africa, 16 species of large mammals went extinct in the past 10,000 years, including nine in historic times. This is in contrast to the general pattern, prior to the emergence of the genus *Homo*, of an extinction rate of about four large mammal species every 100,000 years (McKee 1995).

The ineluctable conclusion is that the growth of our population and the extinction of other species have long been closely related and accentuated with the origins of sedentism and agriculture. Our analysis of contemporary data further demonstrates that population densities and agricultural practices still play a critical role in understanding patterns of extinction.

4.3 Biodiversity and Human Population Density Today

The human population grew past six billion people in1999 and has reached over 6.7 billion by 2009 (US Bureau of the Census 2009a). Our numbers continue to grow such that there will probably be seven billion people by 2013 (US Bureau of the

Census 2009b) and nearly nine billion by 2050 (UNFPA State of World Population 2004). Meanwhile, 11% of known mammal and bird species are threatened (Stork 1997), compounded by immeasurable effects on species yet to be documented by the scientific community. Are these figures directly connected?

There are sound theoretical reasons and considerable evidence suggesting that a close relationship between human population size and biodiversity losses, as in the past, continues today in an alarming manner. Increases in population size and density have caused rapid cultural and ecological changes initiated by human endeavors. Our analysis in this contribution is based on known "threats" to extant species as opposed to documented terminal extinctions, such as those confirmed by our research on the fossil record. Again, the species is a convenient unit of analysis, though genetic and ecosystem biodiversity are also important variables to consider.

Ironically, there are of many examples of human-introduced species that result in biodiversity loss. Globalization of our population, born of necessity as more of us require "unearned resources" from other parts of the world, inevitably globalizes other species, usually considered to be "weed species". Humans may be one such species.

Examples of plant and animal biodiversity loss do not always paint a clear picture of global biodiversity threats. In order to explore a broader view of current trends, McKee et al. (2004) analyzed data on threatened species per nation, comprising critically endangered, endangered, and vulnerable species of mammals and birds from the IUCN Red List (2000). Data from 114 continental nations, excluding exceptionally small nations, was also compiled on human population densities and "species richness" – defined for analysis as the number of known mammal and bird species per unit area. A stepwise multiple regression analysis of log-transformed data defined a statistical model that explained 88% of the variability in current threats to mammal and bird species per country on the basis of just two variables: human population density and species richness (Fig. 4.1). Clearly, "species richness" is not the root cause of the threats – these diverse ecosystems persisted through climatic changes and ecosystem shifts over many thousands of years. That leaves the other variable in the equation, human population density, as the likely culprit leading to globally increased species threats. In essence, a greater concentration of species sets the stage for the human impact to be more devastating.

The human population impact on biodiversity has empirical support from both past and present – it is more than an assumption. On the other hand, correlation does not necessarily mean causation. One must ask if our increased population density is the root cause or a spurious correlation that masks the more direct effects of human behavior. Certainly, one can assume, there must be some effect from what many ecologists now refer to as the "ecological footprint" – the effect each individual or group has in terms of resource consumption (Wackernagel and Rees 1996; Chambers et al. 2000). This is manifested in many ways – fuel consumption, deforestation, fresh water usage, global warming, or even the household dynamics of urban sprawl (Liu et al. 2003). There are direct correlates of the "ecological footprint" with depletion of both renewable and nonrenewable resources. Is this extraction of resources also related to biodiversity losses?

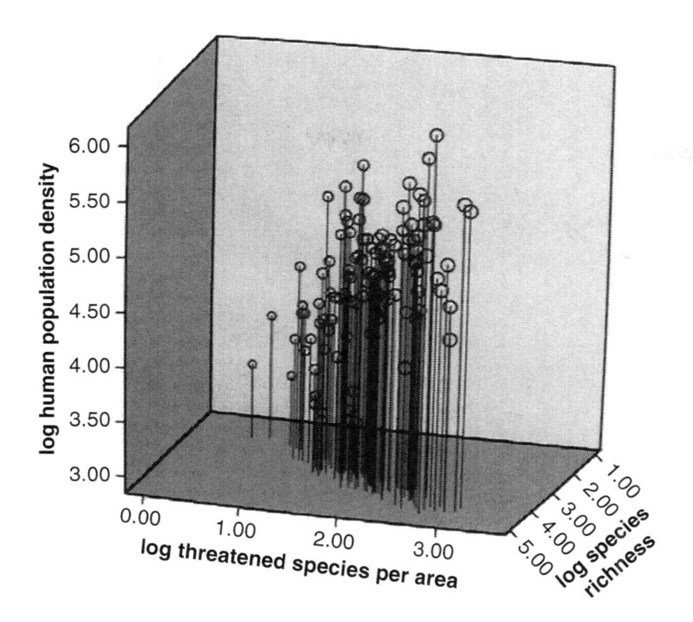

Fig. 4.1 Relationship of threatened species per unit area, population density, and species richness. A multiple regression model, log threatened species per 10^6 km^2 = $-1.534 + (0.691 \times$ log [species richness] + $(0.259 \times$ human population density), accounts for 88% of the variation in species threats

Such questions can be teased from the global data by adding variables to the model and testing hypotheses. One measure of some aspects of the "ecological footprint", for which data are generally available, is per capita Gross National Product (GNP). Previously, McKee (2003) found that whereas there is a strong correlation between species threats and human density, the threat has virtually no correlation with per capita GNP. Figure 4.2 shows the relationship between log-transformed currency-adjusted per capita GNP (Purchasing Power Parity) and the number of species threats for mammals and birds among 101 nations (for which all data were available). The effects of affluence on threatened species originally appeared to be overshadowed by our sheer numbers.

It was somewhat surprising to find virtually no correlation. Kerr and Currie (1995) found a correlation between threatened mammal species and per capita GNP with a different global data set and different methods ($N = 82$ nations), but this was not borne out by our data (which unlike their study excluded small and island nations, perhaps accounting for some of the differences). The reasons behind this counter-intuitive lack of correlation, or the *negative* correlation found by Kerr and Currie, no doubt are complex. But it is clear that the effects of our large population are mediated through a variety of means – just as in the past when the hunting effect was supplanted by the agricultural effect. Kerr and Currie *did*, like us, find a strong population effect on threatened bird species, and other independent tests have also

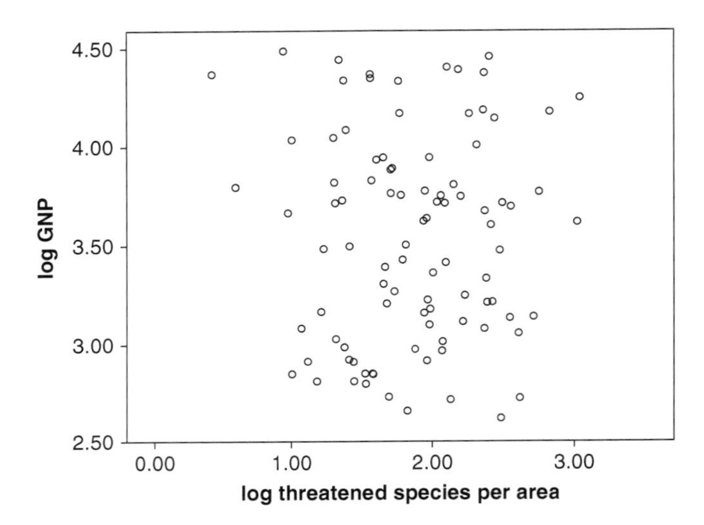

Fig. 4.2 The lack of a close relationship between GNP and threatened species per unit area is evident in this scattergram

highlighted the effects of human population numbers (Kirkland and Ostfeld 1999; Thompson and Jones 1999; Brushares et al. 2001). Large numbers of people in nations rich and poor invariably put pressures on other species that rely upon the same resources.

In order to address these issues further, we reanalyzed the data, looking at GNP per unit area. An interesting, albeit complex, picture emerged. A strong and statistically significant *positive* correlation (Pearson's) between log-transformed GNP and threatened species came into focus ($r^2 = 0.443$, $p < 0.001$). This correlation is evident in the scattergram of Fig. 4.3. By comparison, human population density alone was a slightly lesser predictor of species threats ($r^2 = 0.402$, $p < 0.001$, both variables again log-transformed; Fig. 4.4).

On the other hand, a stepwise multiple regression analysis in which GNP per unit area was added to the variables of the McKee et al. (2004) model left us with the same model: human population density and species richness were the better predictors, to the exclusion of GNP. Part of the reason for this counterintuitive result is that GNP is positively correlated with species richness ($r^2 = 0.445$, $p < 0.001$). Perhaps the high primary productivity of these areas drives diversity as well as economics – but from a statistical perspective, the overlap of GNP and species richness explains some of the same variability in contemporary threats to species of mammals and birds.

Given the archeological association of the origins of agriculture and extinctions of many mammalian species, it is also instructive to look at contemporary correlations between agricultural land use and species threats. We found a statistically significant positive correlation ($r^2 = 0.187$, $p < 0.001$, using log-transformed variables; Fig. 4.5). This correlation is weaker than that of either GNP or

Fig. 4.3 Once GNP is considered per unit area, the relationship to threatened species becomes more apparent. Compare to Fig. 4.2

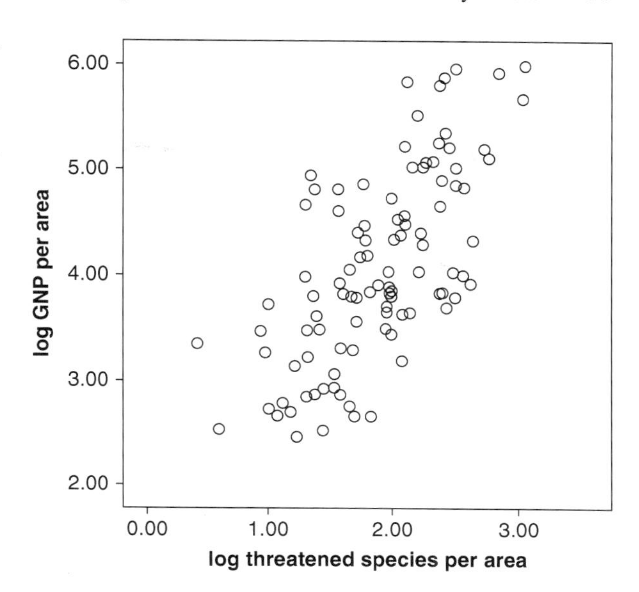

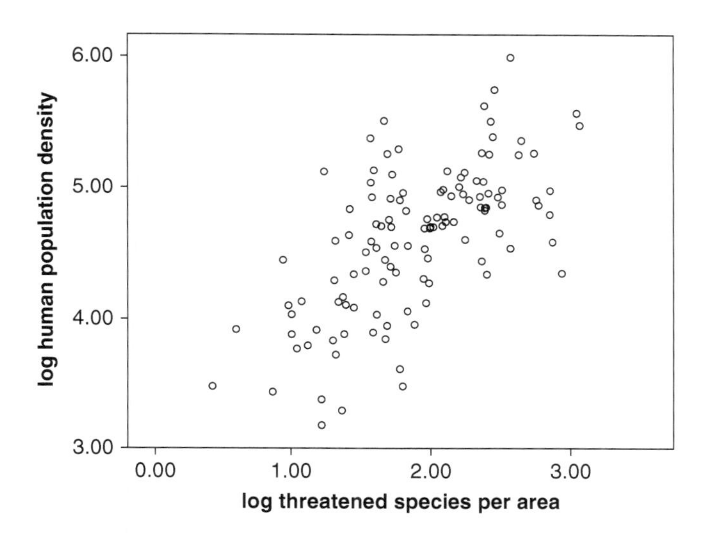

Fig. 4.4 Scattergram of relationship between human population density and threatened species per area

population density. Then again, population density and percentage of land devoted to agriculture are correlated as well ($r^2 = 0.654$, $p < 0.001$). Thus, the question arises as to whether the correlation reflects the direct effect of agriculture usurping the resources of other species or is agriculture simply a mediator of the human population density effect.

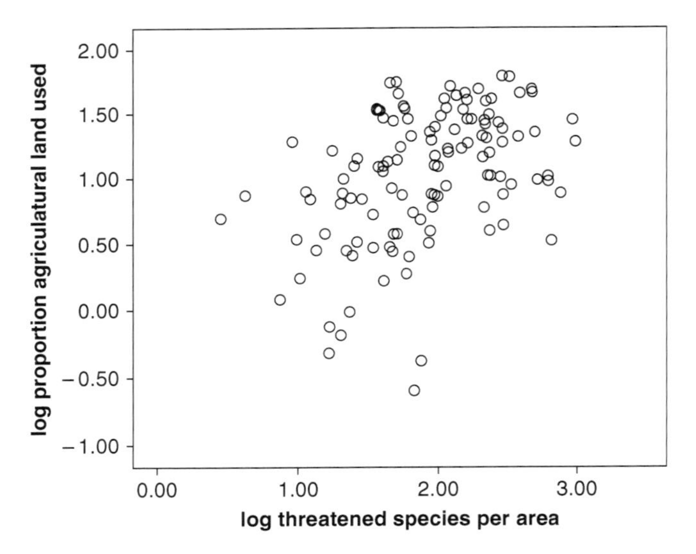

Fig. 4.5 Scattergram of relationship between proportion of active agricultural land use and threatened species per area

Adding agricultural land use into the stepwise multiple regression analysis, we find that it does add a small but statistically significant component to the model predicting nation-by-nation species threats. It explains some of the variability that other variables, including GNP, do not, thereby increasing the predictability of the model from 88% to 89%.

In summary, numerically speaking, once all of the variables used in this analysis are taken into account, species richness, human population density, and agricultural land use are the best combined predictors of threats to species of mammals and birds. GNP – a measure of economic activity that counts residents' income from economic activity abroad, as well as at home – while strongly correlated with species threats, does not add to the predictive ability of the model. These results, combined with long-term observations of the human impact on mammal species, lead us to argue that human population density is a primary cause of biodiversity losses, in a large part mediated by agricultural land use, and thus is a key factor that must be addressed to reduce future threats to Earth's biodiversity.

4.4 Discussion

The results of our analyses bear on debates as to whether human consumption or population density is more relevant in efforts to thwart a mass extinction and its detrimental ecological consequences. Polarized perspectives have emerged. One can take the adamant position of Smail: "*Population stabilization and subsequent reduction is undoubtedly the primary issue facing humanity; all other matters are subordinate*" (2003: 297, italics his). Alternatively, Chambers et al. (2000: 59) exclaim "Don't count the heads – measure the size of their feet". We conclude that such debates are specious, and that a better mantra would be "Count heads, mind your feet". Our research demonstrates that both considerations are relevant, and both must be considered in a comprehensive conservation plan.

For example, one of the complexities not sufficiently addressed in our nation-by-nation statistical analysis is that behavior in one nation can affect biodiversity in another. McKee et al. (2004) noted that Brazil stood out in the analysis as not fitting the trend of greater human population density in species-rich nations leading to biodiversity threats – their threat levels were in excess of those predicted by our model. Such a country may be the exception that proves the rule regarding the importance of global human population growth – economic factors due to population demands in countries with which Brazil does business necessarily influence the rate of habitat destruction and hence the number of threatened species.

Compared to the amount of literature written on conservation to limit biodiversity loss through reduced consumption, nature reserves, and even valuable new ideas such as reconciliation ecology (Rosenzweig 2001, 2003), there is a relative dearth in the wildlife conservation literature on the need to reduce human population density. Here, we want to emphasize the importance of both traditional and novel conservation measures, but concentrate this discussion on population issues as they relate to biodiversity.

Cincotta and Engelman (2000) did present a strong case for the need to address population issues in biologically diverse "hotspots". Our global analysis, which uses country-level data, comes to a similar conclusion – that greater human population has been accumulating in regions associated with higher levels of species endemism – despite our analysis having excluded the islands that comprise many of the hotspots. Clearly, human population growth with the hotspots should be addressed quickly as part of a complete conservation plan. Yet it is our assessment that in order to preserve biodiversity at all levels, we need to go beyond a focus on hotspots, valuable though they may be, to a more global effort in which conservation and human population reduction are both paramount to the survival of the planet.

By way of illustration, the state of Ohio can serve as a case study, for its problems are a microcosm of general global trends. Although it is not a biodiversity hotspot, within its political boundary are at least 175 endangered and threatened species, by state government accounts (Hunt 2005). There is a concentration of public discussion on balancing economic development with preserving Ohio's natural heritage, and many conservation projects have succeeded. On the other hand, of the 2000 or so development projects reviewed each year by the US Fish and Wildlife Service, none have been turned down. Similarly, the Ohio Environmental Protection Agency issued 336 environmental permits for construction and development (e.g., waste water discharge, drilling, and water quality maintenance permits), covering 97.5 ha of wetlands in fiscal year 2004 (Hunt 2005).

Part of the pressure on Ohio for development is the growth of our population, but population issues are rarely considered. There is a general perception in the state, often repeated by various news agencies, that Ohio's human population has remained steady at "about" 11 million people. In very round figures, that may be true, but from 1990 to 2000, Ohio's human population grew 4.7% – by more than half a million people, from roughly 10.8 million to 11.4 million (US Bureau of the Census 2009c). Ohio's rate of population growth is slower than the country as a whole, which grew 13.1% during the same time period. But in a state already saturated with people, it is highly significant. Ohio has the seventh largest population of states in the USA as of 2000, despite being 34th in land area.

So part of the problem is that the general public and policy-makers do not recognize the relevance of our rapidly growing human family. A further component of the problem is that population issues are politically unpopular. There was nearly no mention of population issues in the US presidential election campaigns of 2004 and 2008. This is symptomatic of a larger problem. For example, at the 2002 World Summit on Sustainable Development in Johannesburg, South Africa, population was virtually a taboo subject, despite their goal of reducing the rate of biodiversity loss by 2010.

The complexity of population issues stymies those who should know better from even broaching the subject. Controversial and complex issues concerning human rights and racism, for example, are integral components of the dialog on population growth abatement. But difficulties in addressing such issues should not prevent the conversation from taking place.

Another key component to public diffidence toward population problems is that our rate of growth is slowing. Thus, there is a perception that as developing nations follow the theorized pattern of the demographic transition, our population will naturally peak at 10 or 11 billion, depending on estimates of fertility as the transition occurs (Lutz et al. 2001). There are a number of problems with this logic.

The demographic transition typically involves economic growth and increased consumption, hence increasing the "footprint". In economics, there is no equivalent of the "demographic transition", in which growth slows naturally. It could be argued that affluent societies have more modern technological developments, which represent, at least in theory, progress toward a more efficient, less environmentally draining mode of production. But that is not what we see. For example, as China becomes more industrialized, it is on course to overtake the United States as the most voracious consumer of resources (Favin and Gardner 2006).

Moreover, the underlying assumptions of the demographic transition are not borne out by the data. McKee (2003) argued that many countries did not fit the traditional model. To test this idea, we used our data to compare national growth rates to GNP. Whereas there is indeed a statistically significant correlation ($p < 0.01$), only 52% of the variation in growth rate can be explained by GNP (Fig. 4.5). In other words, we cannot automatically count on the demographic transition through economic development to abate human population growth.

The point we want to make here is that population issues and policy initiatives must move to the top of the political and policy agenda. There is no guarantee that the human population growth will continue to slow naturally through the demographic transition, the alleged economic catalyst of the transition involves increased consumption, and even if our population *does* peak at 10 or 11 billion, that is far too many for sustainability of biodiversity (McKee 2003). Whereas we agree that conservation policies are vitally important to sustaining the ecological health of the planet, they will be all for naught unless we find a way to close the floodgates of human population growth.

The evidence is in the statistics. As we demonstrated with prehistoric and contemporary data, there is a strong and important correlation between human population growth and biodiversity losses. Using our mathematical model (see Fig. 4.1) to forecast future species threats based upon demographic projections per country, all else being equal, it was found that we can expect a 7% increase in the global number of threatened species of mammals and birds by 2020, and a 14% increase by 2050, based upon growth in human numbers alone (McKee et al. 2004). It is difficult to translate these calculations into predicted numbers of extinctions, but as we noted earlier, the very nature of the threat involves extinctions of genetic variability, thereby creating an extinction debt. Without intervention toward abating and halting human population growth, future extinctions are assured.

4.5 Conclusion

Human population growth has resulted in changes in Earth's biodiversity for thousands of years. Competition within global ecosystems has produced evolutionary changes resulting in the rise and fall of species. Although the extinction of a species is a natural event, the frequency of these extinctions is rising at an unprecedented rate in human history. Increases in human population density have initiated drastic changes in land use strategies and heightened levels of migration causing plant and animal displacement and extinction. We stated earlier that "novel adaptations to new modes of existence are necessary components of survival". That is true for our species now.

Increases in human population growth and environmental dominance and manipulation have set the stage for the global mass extinction that has already begun. However, the extinction rate is not the sole indicator of a compromised ecosystem. Species threats are causing a depletion of genetic biodiversity, which puts species in greater risk of extinction since adaptability to altered environments becomes less likely. Ecosystem diversity is also jeopardized by human expansion, thus compounding the threat.

Our analyses have demonstrated that species richness, human population density, and agricultural land use are the best predictors of species threats. Increases in human population density and concomitant lifestyle practices are the primary cause of biodiversity threats. Malthusian principles, although much maligned for two centuries due to the successes of the human enterprise, have snuck up behind us as the biodiversity on which we rely has continued to quietly dwindle to dangerous levels of vulnerability. We need to overcome the public aversion toward identifying and addressing population issues. With the sustainability of global ecosystems under threat, the human family needs to realize that addressing the crisis of overpopulation is in everybody's best interest.

Acknowledgments We would like to thank Richard Cincotta for the invitation to write this chapter as well as his insights that helped guide our analysis.

References

Alroy J (2001) A multispecies overkill simulation of the end-Pleistocene megafaunal mass extinction. Science 292:1893–1896

Armelagos GJ (1990) Health and disease in prehistoric populations in transition. In: Swedlund A (ed) Disease in populations in transition. Bergin and Garvey, New York, pp 127–144

Barrett R, Kuzawa CW, McDade T, Armelagos GJ (1998) Emerging and re-emerging infectious diseases: the third epidemiologic transition. Annu Rev Anthropol 27:247–271

Behrensmeyer AK, Todd NE, Potts R, McBrinn GE (1997) Late Pliocene faunal turnover in the Turkana Basin, Kenya and Ethiopia. Science 278:1589–1594

Brushares JS, Arcese P, Sam MK (2001) Human demography and reserve size predict wildlife extinction in West Africa. Proc R Soc Lond B 269:2473–2478

Chambers N, Simmons C, Wackernagel M (2000) Sharing nature's interest – Ecological footprints as an indicator of sustainability. Earthscan, London

Cincotta RP, Engelman R (2000) Nature's place: human population and the future of biological diversity. Population Action International, Washington, DC

Cowlishaw G (1999) Predicting the pattern of decline of African primate diversity: an extinction debt from historical deforestation. Conserv Biol 13:1183–1193

Favin C, Gardner G (2006) China, India, and the new world order. In: Stark L (ed) State of the World 2006. WW Norton, New York, pp 3–23

Holdaway RN, Jacomb C (2000) Rapid extinction of the moas (Aves: Dinornithiformes): model, test, and implications. Science 287:2250–2254

Hunt S (2005) Habitats in danger? Projects usually not. The Columbus Dispatch 01/01/2005

IUCN (2000) Red list of threatened species. http://www.iucnredlist.org/. Accessed June 2000

Kerr JT, Currie DJ (1995) Effects of human activity on global extinction risk. Conserv Biol 9: 1528–1538

Kirkland GL, Ostfeld RS (1999) Factors influencing variation among states in the number of federally listed mammals in the United States. J Mammal 80:711–719

Klein RG (2000) Human evolution and large mammal extinctions. In: Vrba ES, Schaller GB (eds) Antelopes, deer, and relatives, present and future: fossil record, behavioral ecology, systematics, and conservation. Yale University Press, New Haven, pp 128–139

Larsen CS (1995) Biological changes in human populations with agriculture. Annu Rev Anthropol 24:185–213

Liu J, Daily GC, Ehrlich PR, Luck GW (2003) Effects of household dynamics on resource consumption and biodiversity. Nature 421:530–533

Lutz W, Sanderson W, Scherbov S (2001) The end of world population growth. Nature 412:543–545

McKee JK (1995) Turnover patterns and species longevity of large mammals from the late Pliocene and Pleistocene of southern Africa: a comparison of simulated and empirical data. J Theor Biol 172:141–147

McKee JK (2001) Faunal turnover rates and mammalian biodiversity of the Late Pliocene and Pleistocene of eastern Africa. Paleobiology 27:500–511

McKee JK (2003) Sparing nature – the conflict between human population growth and Earth's biodiversity. Rutgers University Press, Piscataway

McKee JK, Sciulli PW, Fooce CD, Waite TA (2004) Forecasting global biodiversity threats associated with human population growth. Biol Conserv 115:161–164

Mlot C (1995) Biological surveys in Hawaii, taking inventory of a biological hot spot. Science 269:322–323

Pimm SL, Russell GJ, Gittleman JL, Brooks TM (1995) The future of biodiversity. Science 269: 247–250

Redman CL (1999) Human impact on ancient environments. The University of Arizona Press, Tucson, Ariz

Roberts C, Manchester K (1997) The archaeology of disease. Cornell University Press, Ithaca, New York

Rosenzweig ML (2001) Loss of speciation rate will impoverish future diversity. Proc Natl Acad Sci 98:5404–5410

Rosenzweig ML (2003) Win-win ecology: how the earth's species can survive in the midst of human enterprise. Oxford University Press, Oxford

Smail JK (2003) Remembering Malthus III: implementing a global population reduction. Am J Phys Anthropol 122:295–300

Stork NE (1997) Measuring global biodiversity and its decline. In: Reaka-Kudla ML, Wilson DE, Wilson EO (eds) Biodiversity II. Joseph Henry, Washington, DC, pp 41–68

Thompson K, Jones A (1999) Human population density and prediction of local plant extinctions in Britain. Conserv Biol 13:185–190

UNFPA State of World Population (2004) http://www.unfpa.org/swp/

US Bureau of the Census (2009a) US and World population clocks. http://www.census.gov/main/www/popclock.html. Accessed 1 February 2009

US Bureau of the Census (2009b) International Data Base. http://www.census.gov/ipc/www/idb/worldpopgraph.html. Accessed 1 February 2009

US Bureau of the Census (2009c) American Factfinder. http://factfinder.census.gov/home/saff/main.html?_lang=en. Accessed 1 February 2009

Wackernagel M, Rees W (1996) Our ecological footprint – reducing human impact on the earth. New Society, Gabriola Island, British Columbia

Chapter 5
The Biological Diversity that Is Humanly Possible: Three Models Relevant to Human Population's Relationship with Native Species

Richard P. Cincotta

5.1 Introduction

Almost universally, ecologists agree that the remarkable geographical diffusion of *Homo sapiens*, followed more recently by the species' even more remarkable numerical increase, have played significant roles in bringing about rapid and ongoing declines in *biological diversity* (biodiversity) – a term that encompasses our planet's vast array of biotic species, genetically distinct subspecies and varieties, local breeding populations and the ecological systems (ecosystems) in which they live, reproduce, and evolve. Despite this firm consensus, or perhaps because it is so firm, no major cohesive body of testable theory has yet emerged from within the ecological sciences that specifically focuses on human demographic variables and their relationships to changes in biodiversity. On this set of relationships ecologists are informed, by and large, by population biology's "first principals" – predictions of the most fundamental theories derived to explain the dynamics of interacting populations (see May 1973) and by piecing together the conclusions of empirical research from several applied ecological fields, particularly conservation biology, landscape ecology, and community ecology. To scientists from other disciplines, this hodgepodge of evidence may appear as a shaky foundation for such overwhelming consensus. Thus, it is fair to ask: What existing theories, from ecology or the social sciences, could be modified to explain observed dynamics of interactions between human population variables and the population biology of other species, to make predictions under changing conditions, and to test these expectations?

In this essay, I discuss three possibilities and how they are, or can be made, germane to the study of human–biodiversity interactions. Each model, when applied to empirical research, offers a response to a fundamental question about these interactions. The first is a contentious question about population size: Can human numbers by themselves affect biological diversity? A set of empirical models that describe the relationship between body mass and expected population density for mammalian herbivores and carnivores are applied to *Homo sapiens*, and its implications are briefly discussed. The second question is about population growth: Can local changes in human population density affect biological diversity?

R.P. Cincotta and L.J. Gorenflo (eds.), *Human Population: Its Influences on Biological Diversity*, Ecological Studies 214,
DOI 10.1007/978-3-642-16707-2_5,

The model that follows explores the dynamic demand for inputs in the process of density-dependent intensification of agricultural production systems, drawing its graphical form from a conceptual model in rural sociology. The third question is about human distribution: Can the location of human settlement affect biological diversity? The model that delves into this question draws from the perspective of island biogeography in the ecological sciences. By identifying the factors that mediate the continuity of a species in an isolated habitat, the third model suggests how human population density and activity could influence whether a native species is present or absent.

5.2 Can Human Numbers by Themselves Increase the Risk of Species Loss?

5.2.1 Allometric Relationships

Biologists have found numerous significant statistical associations relating anatomical, physiological, morphological, and ecological variables ($\boldsymbol{Y}_i$) to the live body weights of individuals in comparable groups of animal species ($\boldsymbol{W}$). Known as *allometric* relationships, these associations take the form of $\boldsymbol{Y}_i = a\boldsymbol{W}^b$, where b determines the path of the curvilinear relationship and a controls its scale. Log-transformed, these relations assume a linear form, $\mathrm{Log}(\boldsymbol{Y}_i) = b\mathrm{Log}(\boldsymbol{W}) + \mathrm{Log}(a)$, and thus linear regression can be used to determine statistically whether this relationship holds between animal body weight and a hypothetically associated variable.

Physioecologists have noted that, in comparisons of evolutionarily related groups of animal species, basal metabolic rates (BMR) vary positively with average adult body mass. There are, however, efficiencies in scale. Heat tends to be conserved in animals of larger body mass, an effect principally due to the geometric fact that, among animals of similar anatomical design, surface area (from which heat escapes) per volume decreases as body volume increases. Because larger animals can afford larger gut volumes and slower rates at which their food passes through their gut, larger animals tend to experience greater digestive efficiencies, as well (Van Soest 1982). Thus, energy requirements *per unit body mass* vary inversely with species' body mass. Empirically, for any related group of similar animal species of varying sizes, BMR tends to increase as a function of $\boldsymbol{W}^{-0.75}$.

5.2.2 Allometric Expectations of Population Density

Recognizing that species with larger body mass have greater energy requirements and generally must range farther for food, Robert Henry Peters (1983) used this function to relate estimates of abundance over large areas of relatively undisturbed native habitat to species' average body sizes. By regressing the natural-log transformed values of abundance and average adult body mass, Peters determined the following relationships predicting densities (animals km^{-2}) of terrestrial mammalian herbivores (grazers and browsers), D_H, and carnivores, D_C:

$$D_H = 103\,W^{-0.93}, \tag{5.1}$$

$$D_C = 15\,W^{-1.16}. \tag{5.2}$$

5.2.2.1 Expected Density of Preagricultural Humans

Because they are based on near-equilibrium animal densities (among a full compliment of prey or forage, competitors, and predators), in relatively undisturbed native habitat, these functions are clearly not directly applicable to *Homo sapiens* practicing agriculture. These equations can, however, be used to estimate how many preagricultural humans could have been sustained at near-equilibrium densities in ecological systems if our species had remained, in effect, just another big, terrestrial mammal: a primate that subsisted, without the benefits of cultivation, as a granivorous (seed-eating), frugivorous (fruit-eating) species with a predilection, and acquired skills and tools, for carnivory – whether by hunting or scavenging for meat.

Because animal-density data were gathered primarily from tropical and temperate ecosystems, and I have not corrected for low productivity desert or arctic ecosystems, the global prediction for preagricultural human densities is likely to be high. However, I have not allowed for foraging and hunting in nonterrestrial ecosystems, such as coastal fishing, which has served as a primary means of obtaining protein and food energy in near-arctic, island, and other coastal environments, which often supported human populations at higher densities than contemporary inland settlements (Renouf 1984). For an herbivorous mammal with the size of *Homo sapiens*, averaging roughly 65 kg, the appropriate equation (5.1) predicts 2.1 individuals km^{-2}. For a carnivore of the same weight, the appropriate equation (5.2) predicts 0.12 individuals km^{-2} (also see Cohen 1995). Preagricultural human diets, however, probably fell in-between carnivorous and herbivorous diets. A liberal estimate of the average population density that our species would likely have maintained *without agriculture* is around 1.0 km^{-2} individuals. In 2005, world population density was estimated at 48 km^{-2} individuals (United Nations Population Division 2005). UN models project that by 2050 human density will likely fall within the range of 50–72 km^{-2}.

Because of hospitable climate, fertile soils, and abundant fresh water, and the local evolution of systems of resource distribution, financing and trade that encouraged urban growth, and relatively high rural densities, in some countries humans have reached densities that, in comparison to the world average, are extraordinarily high. In the year 2005, for example, Bangladesh had reached a density of about 985 km^{-2}, while population density in the Netherlands was about 392 km^{-2} the same year (United Nations Population Division 2005).

5.2.2.2 Historical Human Densities

If preagricultural humans at densities ranging from around 1.0 to about 1.5 km^{-2} were to exploit every corner of Earth's habitable terrestrial surface, which has been estimated at about 130 million km^2 (FAO 2000; Hannah et al. 1994), the world could conceivably support from 130 million to around 200 million preagricultural humans. According to several estimates, world population surpassed 130 million in the early years of the Roman Empire (before 4000 BCE, some 10,000 years after the emergence of agriculture in the Mediterranean Basin) and reached 200 million around 200 CE (Biraben 1979; Livi-Bacci 1992; Cohen 1995). Most of this population was agricultural, living clustered in relatively high densities in the Mediterranean Region, in the fertile flood plains of central Europe and Asia, and in the highlands of tropical America and Africa.

Current world population, estimated by the UN Population Division at about 7.0 billion people in mid-2011, is roughly 35–55 times what preagricultural humans would likely have achieved, according to this conservative calculation.

This hypothetical range of human densities, around 1.0–1.5 km^{-2}, represents a first approximation of the limits of preagricultural human populations in many natural ecosystems. As local human population density grew well beyond this range, preagricultural members of our species were compelled to physically alter landscapes, control nonhuman competitors and predators, modify and husband sources of protein, digestible nutrients and food energy (develop crops and livestock), or find food sources outside the terrestrial realm (such as within littoral and marine ecosystems) in order to ensure their own survival and reproduction. Thus, even putting aside the substantial biological impacts of modern humanity's consumption patterns and related industrial wastes, the acute concerns of overhunting of terrestrial native species and overexploitation of fisheries, and the threats associated with invasive species, there is little doubt that, on its own, the scale of human numbers represents a fundamental challenge to the integrity of the current (remnant) array of biological species.

5.3 Can Local Changes in Human Population Density Increase the Risk of Species Loss?

5.3.1 Density-Dependent Agricultural Intensification

As the starting point for her thesis positing human population growth as a root cause of agricultural intensification, sociologist Ester Boserup (1965) chronicled the evolution of agricultural production, from primitive farming systems that relied on *long fallows* to more intensively farmed *short-fallow* systems. Households employing long-fallow production systems – including shifting cultivation (swidden, or slash-and-burn farming), nomadic grazing, and long-rotation dryland farming – relied on the lengthy natural processes of vegetation regrowth, plant reproduction, and secondary plant succession to replenish biomass, and on decades-long cycles of decay and mineralization to replenish the soil's available nutrients and organic matter.

Boserup observed that increased human population density among producers – those with access to productive land and rights to harvest its products – led to changes in large-scale patterns of land use and relatively rapid increases in productivity (kg of product ha^{-1}). Producers who inherited a smaller farm plot or were constrained to a smaller grazing allotment, or who faced greater competition for resources on communal land, were forced into shorter fallow cycles and larger output per hectare just to maintain a subsistence level of consumption.

To increase average output per unit area, producers are driven to expend more effort, adopt more closely managed modes of production, and use technologies that facilitate more rapid cycling, greater plant productivity, and greater energy and material efficiencies. To protect their investments, short-fallow-system farmers and sedentary graziers claimed and fenced land (Boserup 1965), developing traditional systems of property rights that, in many cases, have been incorporated into state legal systems. The expectations of Boserup's theory contest Malthus's earlier assumption that agriculture production typically grows linearly overtime and would ultimately be outpaced by the needs of exponentially growing human populations, promoting constraints on further increases of human population, including famine, disease, and warfare. Applied to local systems in the contemporary world, Boserup's model of agricultural change has, so far, rewarded social scientists with considerably more explanatory power than that of Malthus.

However, several limitations of Boserup's conceptual model inhibit its wider application to agricultural economics. The first is Boserup's reluctance to account for large-scale constraints to continuous intensification. The most obvious is the diminishing availability of renewable freshwater resources in some areas, which has prevented some producers from shifting to more intensive production systems. In cases where the irrigated agriculture's demand for water has been politically outcompeted or outpriced by urban and industrial water demands – as in parts of the western Great Plains of the United States – farmers have sometimes sold off portions of their water rights and shifted to less water-intense, longer-fallow systems, such as low-intensity grazing (rangeland beef production) and dryland grain production (Gollehon and Quinby 2000). In addition, Boserup's theory does not come to grips with the powerful production incentives and disincentives that have been generated by agricultural integration into national political economies, into regional markets, and into the global system of international trade, tariffs, and food aid. And finally, and perhaps most importantly, Boserup's conceptual model fails to explain the outcomes of producer depopulation: observed increases in the use of seasonal extrinsic labor, capital equipment, and technology that are substituted when producer population density declines, which is often the case when farmer family members (particularly young adults) migrate to urban areas, when family members take off-farm employment to diversify income, or when farms are consolidated (Synapse Research and Bob Hudson Consulting 2005). Population decline among farm producers is commonplace in industrial countries (e.g., in the

Great Plains region of the United States and in Australia) and is likely to occur in the near future, locally in some Asian countries, as the agricultural sector mechanizes and wages increase in urban manufacturing and service industries.

5.3.2 A Graphic Model of Density-Dependent Intensification

Population density-driven intensification can be modeled (Fig. 5.1) by imagining a community of agricultural producers that live in an environment in which individual producer households can shift between a set of discrete production systems, **S**: {$\mathbf{S}_i$, $\mathbf{S}_{i+1}$, . . . , $\mathbf{S}_j$}. The labor efficiency (**E**) of a producer in each system varies monotonically with the number of producers per arable land surface area, and therefore in each system, there is a unique maximum efficiency ($\boldsymbol{e}_i$) at an optimal producer population density ($\boldsymbol{d}_i$). To be a viable production system, its efficiency must peak above a threshold of producer solvency ($\boldsymbol{e}_s$). As the producer density grows well beyond its system's optimum, the need to remain viable forces producers to switch to another system. In practice, however, as density-dependent efficiencies decline, producers tend to compensate with increased labor intensity and energy use, and then to modify the system by adding infrastructure and technological inputs (and thus alter the system's efficiency function). Ultimately, increased producer density triggers system switching when households can manage switching's transition costs (costs to build new infrastructure, to acquire new technologies, to learn new skills, and to maintain institutions).

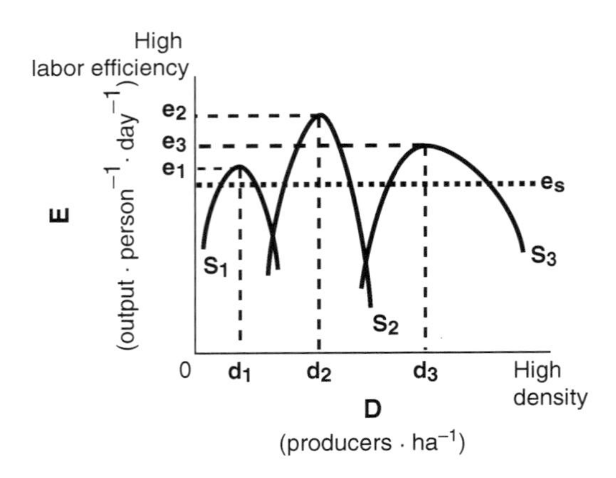

Fig. 5.1 A graphic interpretation of Boserup's theory of density-dependent agricultural intensification. In this model, the labor efficiency (**E**) of each possible viable agricultural production system ($\boldsymbol{S}_i$) varies monotonically as a function of the density of producers (**D**). To be viable, the maximum efficiency of a system ($\boldsymbol{e}_i$), which occurs at a producer population density ($\boldsymbol{d}_i$) must peak above a threshold of producer solvency ($\boldsymbol{e}_s$). As their density increases, agricultural producers are driven to increase effort and energy use to cope with a decline in labor efficiency, and are ultimately compelled to switch between discrete systems

For ecologists, this model's most salient dynamics are implied by the phenomena of forced density-dependent switching and system modification. Historically, both processes have tended to alter ecosystem structure and biogeochemical cycling, and they do not necessarily yield an agricultural production system that is more efficient (a higher peak efficiency) than the prior system.

However, unlike Boserup's conceptual theory of density-dependent intensification, this model allows for producer populations to decline below optimum density. In such a case, producers can apply more effort per person (typically in the form of hired labor) or more energy, choose to switch to a lower order system, or modify their current system, protecting their investments in infrastructure, technology, and skills – options that correspond to the realities of agricultural production.

5.3.3 *Model Expectations*

5.3.3.1 Responses in the Range and Domain of the Model

The model's composite functional form provides several research expectations, most of which are consistent with Boserup's conceptual model. Accordingly, growth among the population of producers promotes continuous infusions of labor (number of workers), effort (time spent by workers), capital and technological inputs, and periodically, discrete conversions from one agro-ecological system to another with a larger per-area investment in capital equipment, infrastructure, and landscape-level conversions (irrigation, terracing, grading, and other types of soil movement and enhancement).

What should happen if producers are economically or technologically constrained from shifting to a more energy-intense production system? Because this model restricts responses to the domain of population density, households are expected to constrain growth in population density in order to maintain a constant level of production per person. Possible "real life" responses consistent with this prediction include: household efforts to limit increases in family size by limiting childbirth; efforts to limit the number of household members who are dependent on farm production by encouraging family members to migrate or seek off-farm employment; or the abandonment of farming, where households sell-off property or rent it to other producers, in order to seek income-generating employment through other activities.

What response does the model suggest when the population of producers declines? Because each system's intensity isocline is assumed to be parabolic, the model's predictions of responses to reductions in average household size or numbers of households are ambiguous. A response in terms of changes in the range of energy inputs depends on where, along a system's intensity isocline, the current population is situated. Therefore, to make up for a suboptimal population, a household is expected to add labor and technology rather than shift to a less-energy intense system. Dramatic reductions in the producer population, however, could force a shift to a less intensive system.

Responses along the population domain are also possible responses to declining household size or a decreasing number of households. Household members' perceptions that their numbers currently are, or will later drop, below the optimal compliment for their production system could drive households to add additional children – a hypothesis that has previously been posed in labor-intensive agricultural systems where wealth flows from children to parents (Caldwell 1976). Increasing producer density to move closer to a perceived optimum level also could be achieved by inviting relatives to share in the venture, adding wives in polygamous societies, or inviting nonfamily members into cooperative relationships.

5.3.3.2 Expected Effects on Biological Diversity

While the model of density-dependent intensification presented here does not explicitly provide expectations for researchers studying human population's interaction with biological diversity, at least four biodiversity-relevant hypotheses can be extended from it.

1. *Nutrient theft hypothesis.* As Ester Boserup suggested in her work on density-dependent intensification (1965, p. 52), shortened fallows increase producers' demands for nutrients, organic matter, and water, which are met by depriving nearby ecosystems and their species of these inputs. Grazing animals tend to move organic material and nitrogen by foraging on rangeland and pastures and defecating in areas where they are bedded or sheltered (Turner 1998). Because applications of irrigation water, concentrated animal waste, and industrial fertilizers increase the chances of runoff and leaching, other ecosystems inadvertently receive unusually high concentrations of these inputs and are altered by them. This rapid transport, depletion of some ecosystems, and nutrient pollution of others undermines the viability of populations of native species.
2. *Successional retreat hypothesis.* Increasingly, rapid harvesting and heavy grazing favor plant species that mature and produce propagules quickly (i.e., early successional species, such as annuals and pioneer species) and disfavor animal species that rely on late-successional plant species for their microenvironment, the protective cover they supply, the forage they produce or nutrients they cycle, or the reproductive habitat they provide. The high frequency of soil disturbance typical of short-fallow systems facilitates the establishment of light-seeded plant species and their animal associates. Thus, human-dominated ecosystems tend to support a large component of annual and shade-intolerant plants, and highly mobile exotics.
3. *Anthropogenic creep hypothesis.* Investments in capital and labor, the cultivation of crops and the husbandry of domestic animal species, the dissemination and deposition of anthropogenic wastes, and the construction of infrastructure and human shelter weaken the competitive abilities of native species and promote unusually large populations of scavenging omnivorous species. These gradual (temporally and spatially "creeping") changes also provide novel niches that facilitate the establishment and reproduction of nonnative species and the

overabundance of natives that have adapted to human-dominated ecosystems. Human material transport systems increase the probability that nonnative species or their propagules will be imported and become established (Rapoport 1993).
4. *Institutional bias hypothesis*. As human density increases, native ecosystems and their constituent species become more difficult to protect. Simultaneously, increases in human population density generally stimulate intensive use and higher land values. To protect or manage land and water resources, humans have established *institutions* (agreements, controls, incentives, and disincentives). Because institutions generally promote the well-being of those who devise their rules, the conditions that promote native species richness are likely to lose out in conflicts between users and institutional regulation of land use (Cincotta and Engelman 1997, 2000).

Each of these hypotheses predicts that as ecosystems become more intensively occupied by humans, there is an increased risk that native-species richness (the number of native species present) will decline. However, as the previous discussion suggests, whereas increases in the population density of producers are associated with long-term shifts in production intensity, these shifts tend to be discontinuous (featuring abrupt jumps to higher levels of inputs). Therefore, researchers could observe discontinuities in impacts on ecosystems and native-species richness.

While the theory of density-dependent intensification does not yet provide a direct analytical relationship between human density and the viability of native-species populations, these four hypotheses provide some general theoretical guidance for applied researchers studying the dynamics of native species. And, because human population density is likely to increase in most regions of the world for decades into the foreseeable future (United Nations Population Division 2005) more theoretical research should be conducted to relate human population density to the manner in which human-altered ecosystems function, and its influence on native-species richness.

5.4 Can the Location of Human Settlement Increase the Risk of Species Loss?

5.4.1 Modeling the Persistence of Native Species

The following model builds on the thesis that human influences on populations of native species could be more clearly understood by decomposing the population dynamics of a native species, and then working to understand human influence on each component. In this model, I assume that the overall probability of the presence of a species is composed of three probabilistic components: *intrinsic viability*, *recruitment viability*, and *reestablishment probability*. Intrinsic viability is the population's probability of long-term survival in a discrete habitat, discounting any demographic and genetic contributions from migrants – as if this habitat were undisturbed and isolated from all gene flow for a period of time relevant to long-term management. Factors influencing intrinsic viability include the area and species-specific quality of the habitat, the frequency of population-perturbing events (e.g., severe weather and disease outbreaks) as well as other factors affecting intrinsic rates of birth, death, and loss throughout migration.

Recruitment viability is the additional population viability rendered by in-migrants from nearby habitats in the protected-area network. Factors affecting recruitment are often species-specific: related to its capacity to migrate – to walk, fly, swim or be disseminated by biotic and abiotic vectors during the various stages of its life history. The distance to other habitats that support the species, and each habitat's location in a web of habitats and interhabitat corridors, are also critical factors affecting the probability that recruits will join the population.

Reestablishment probability, which is influenced by most of the conditions that affect recruitment viability, is the probability that the available pool of potential in-migrants could reestablish a population should the resident population die out. Thus, the long-term persistence of a species in a discrete habitat or protected area can be described by the decomposition:

$$E_{ij} = \varpi_{ij} + \mu_{ik}(1 - \varpi_{ij}) + t_{ik}[1 - \varpi_{ij} - \mu_{ik}(1 - \varpi_{ij})] \quad (5.3)$$

where E_{ij} is the long-term probability (i.e., $0 \geq E \leq 1.0$) that a population of a species, i, will exist in an area, j. This probability is, in turn, the sum of three component probabilities: $\overline{\omega}_{ij}$ is the probability that this population would remain viable due to its intrinsic viability; μ_{in} represents the contribution from recruitment viability resulting from in-migration from other habitats, k, in the protected-area network; and τ is the probability that migrants could reestablish a completely decimated population.

5.4.2 Species Persistence and the Influence of Human Settlement

This formulation suggests that there are several direct and indirect avenues through which human settlement affects the long-term dynamics of species composition in protected areas and that human settlement need not occur near primary habitat or inside the recognized boundaries of these areas to be of influence.

For researchers studying the relationship between human population factors and the viability of a native species, the model suggests two possible approaches (a) to determine the sensitivity of the total viability of a population of a native species to the density of human settlement, speculating on the contribution of each component to this viability, or (b) to isolate and determine a component of total viability. Both approaches could lend valuable information to organizations working to plan the establishment of and to manage habitat reserves.

While most studies focus on the proximate causes of decline in breeding populations – habitat fragmentation and loss, biological invasion, pollution, introduced species and diseases, overharvesting, and climate change (Soulé 1991) – rather than the ultimate effects of varying levels of human population density and growth, and patterns of human settlement and movement, there are a few exceptions among the ecological literature relating human population and species viability. Research by Hoare and Du Toit (1999; also see Parker and Graham 1989) is perhaps the most explicit example of an association between human population density and the local viability of a species. The study indicated that above a human density threshold, which the researchers identify at between 15 and 20 humans km^{-2}, elephants move out of southern African shrubland and dry forest. Gorenflo (in this volume) has shown that human population density in semiarid ecosystems in northwestern Mexico and the southwestern United States can indicate, to a reasonable probability, the absence and long-term viability of native species populations. And, where assessed, there is evidence that human settlement density – beyond a very low threshold – promotes a rapid decline in native vascular plant species as a proportion of the total number of species (Rapoport 1993).

Reestablishment probabilities and the length of time to population stability have been studied on very small islands where arthropod species have been experimentally manipulated (Simberloff and Wilson 1969) and by studying the new volcanic islands and zones biotically sterilized by volcanic eruption, such as the resettlement of Mount St. Helen, in the Sierra Nevada range of the northwestern United States (Franklin and MacMahon 2000). However, studying either of these parameters under conditions where production systems and modes of settlement of human populations impede the flow of migrating individuals is likely to be a more difficult, albeit a more rewarding, challenge.

5.5 Conclusions

Besides serving as a basis for debate on human–biodiversity impacts, and possible guides for research on the interactions between human population and biodiversity, some limited conclusions can be drawn from the models introduced and the literature reviewed in this essay.

Application of the allometric model (5.1 and 5.2) to *Homo sapiens* suggests that, because of the relatively large average body weight of adults of our species, the anthropogenic activities in most subsistence production systems (even absent major technologies) are likely to be sufficient to significantly perturb constituent populations of plants and animals in surrounding native ecosystems. According to this analysis, researchers should expect hunter-gathering in many natural ecosystems of densities of even less than 2.0 humans km^{-2} to require management, particularly where humans are relatively recent settlers and have modern hunting technologies at their disposal.

The second model outlined in this essay portrays the increase in production intensity and the successive shifts to more productive systems that have been associated with increases in human population density globally. This model (see Fig. 5.1) reflects the results of studies demonstrating that, as producers become more populous on a fixed area of land, they are motivated to make continuous and discontinuous shifts in *inputs* – including the energy expended per unit area by labor and machinery and the monetary energy-equivalent invested in nutrient and other chemical inputs, in management, in equipment, and in infrastructure. While this model does not functionally link species richness to human population density, it yields insights consistent with four conceptual models that I have very briefly described (a) the nutrient theft hypothesis; (b) the successional retreat hypothesis; (c) the anthropogenic creep hypothesis; (d) and the institutional bias hypothesis. These conceptual models could – with more elaboration and with exploration and testing – yield mechanistic functional forms that could help planners minimize the impacts of development.

The third model presented in this essay (5.3) decomposes the persistence of a native species into its components, each of which could be used to measure impact from varying densities of human settlement and activity. Persistence is assumed to be a function of (a) *intrinsic viability*, the reproduction and survival of a species in a habitat patch of a certain size; (b) *recruitment viability*, the survival of migrants moving between habitat patches; and (c) *reestablishment probability*, the chances of reestablishing a wholly decimated population. Clearly, human settlement, large agricultural plantings (typical of commercial agriculture), and interconnecting roadways that fragment natural habitat affect each of these parameters. In addition, mobility, size, diet, and other life history constraints create major differences in species' sensitivity to human settlement. Given the growing isolation of many wetlands and reserves, particularly in the eastern United States, much more could be done either experimentally or with historical data to assess human settlement's long-term effects on the viability of populations of native species.

The models presented in this essay illustrate varying theoretical aspects of human population's contribution to Earth's ongoing, inexorable shift from a varied array of native ecosystems to a more homogeneous collection of human-dominated ecosystems. In these latter types of ecosystems – where environmental improvement is generally framed by measures of human well-being rather than by considerations of other living components – the selective forces of interacting demographics, economics, and politics act more strongly on populations of species than does natural selection itself. Thus, it would seem to be the principal challenge for theoretical and empirical researchers who are interested in maintaining a large remnant of our natural heritage to better understand both the power and pervasiveness of these anthropogenic forces on native-species richness and on the world's supply of biological diversity.

Acknowledgment I thank Robert Engelman and Larry Gorenflo for editing and commenting on earlier drafts of this essay. Problems that remain are of my own doing.

References

Biraben J-N (1979) Essai sur l'evolutiondu nombre des hommes. Population (Paris) 34(1):3–25, Reprinted in: Livi-Bacci M (1992) A concise history of world population. Blackwell, Cambridge, MA

Boserup E (1965) The conditions of agricultural growth: the economics of agrarian change under population pressure. Aldine, Chicago

Caldwell JC (1976) Toward a restatement of demographic transition theory. Popul Dev Rev 2(3/4):321–366

Cincotta RP, Engelman R (1997) Population and rapid change. Occasional paper, 3. Population Action International, Washington, DC

Cincotta RP, Engelman R (2000) Nature's place: human population and the future of biological diversity. Population Action International, Washington, DC

Cohen JE (1995) How many people can the earth support? W.W. Norton, New York

FAO (Food and Agricultural Organization of the United Nations) (2000) FAOSTAT database. FAO, Rome

Franklin JF, MacMahon JA (2000) Messages from a mountain. Science 288:1183–1185

Gollehon N, Quinby W (2000) Irrigation in the American West: area, water and economic activity. Water Resour Dev 16(2):187–195

Hannah L, Lohse D, Hutchinson C, Carr JL, Lankerani A (1994) A preliminary inventory of human disturbance of world ecosystems. Ambio 23(4/5):246–250

Hoare R, Du Toit J (1999) Coexistence between people and elephants in African savannas. Conserv Biol 15(3):633–639

Livi-Bacci M (1992) A concise history of world population. Blackwell, Cambridge, MA

May RM (1973) Stability and complexity in model ecosystems. Princeton University Press, Princeton

Parker ISC, Graham AD (1989) Elephant decline: downward trends in African elephant distribution and numbers (parts I and II). Int J Environ Stud 34(13–26):287–305

Peters RH (1983) The ecological implications of body size. Cambridge University Press, New York

Rapoport EH (1993) The process of plant colonization in small settlements and large cities. In: McDonnell MJ, Pickett STA (eds) Humans as components of ecosystems. Springer, Berlin, pp 190–207

Renouf MAP (1984) Northern coastal hunter-fishers: an archaeological model. World Archaeol 16(1):18–27

Simberloff DS, Wilson EO (1969) Experimental zoogeography of islands: the colonization of empty islands. Ecology 50:278–296

Soulé ME (1991) Conservation: tactics for a constant crisis. Science 253:744–750

Synapse Research & Consulting, Bob Hudson Consulting (2005) Australian farm sector demography: analysis of current trends and future farm policy implications, August 2005. Australian Farm Institute, Surry Hills

Turner MD (1998) Long-term effects of daily grazing orbits on nutrient availability in Sahelian West Africa: I. Gradients in the chemical composition of rangeland soils and vegetation. J Biogeogr 25(4):683–694

United Nations Population Division (2005) World population prospects: the 2004 revision. United Nations, New York

Van Soest P (1982) Nutritional ecology of the ruminant. O&B Books, Corvallis, OR

Chapter 6
Biodiversity on the Urban Landscape

Katalin Szlavecz, Paige Warren, and Steward Pickett

6.1 Introduction

Expanding urbanization is one of the leading types of land use change today. In 2005, 49.2% of the world population lived in cities, and this number is expected to reach 60% by 2030 (United Nations Population Fund 2007). Urban population in the US is already above 70%, while in the developed countries in general, urban population accounts for more than 80% of national totals (Table 6.1). Even more important for biodiversity is the rate of change of urban and suburban land covers. On a regional basis, the rate of land conversion to urban uses, in the broad sense, often exceeds the population growth in that same region. For example, from 1982 to 1997, developed land (according to the U.S. Department of Agriculture's Natural Resource Conservation Service, land is classified as one of the several land cover/use categories, either urban or other built-up areas, or rural transportation land) in the 48 contiguous United States increased by 34%, while during the same period, population increased by only 15% (USDA NRCS 2001; US Census Bureau 2000). Conversion of land from agricultural and wild categories to the general category of urbanized uses was thus more than twice as fast as population growth for the same 15-year period. Such changes are quite significant for biodiversity (Forester and Machlis 1996).

Cities – a term we will often use as shorthand for the broader array of urbanized areas, from central business districts to old residential areas, to commercial and industrial sites, to new suburbs, as well as the new edge cities and exurban fringe – affect biodiversity because they present unique habitats. Cities are densely populated, highly modified systems resulting from destruction, alteration, and fragmentation of the original wildland or older rural habitats and from creation of new habitat types. Built structures and impervious surfaces make up a large percentage of urban land cover, while remnants of original habitats may still exist. In addition, "volunteer" or semiwildlands are important in some urban areas. Urban landscape is a patchwork of many land uses which, along with altered hydrology (Paul and Meyer 2001; Groffman et al. 2003) and climate (Botkin and Beveridge 1997; Brazel et al. 2000), profoundly affect biodiversity at all spatial scales (Sukopp and Starfinger 1999). Many elements of this landscape are heavily managed by

R.P. Cincotta and L.J. Gorenflo (eds.), *Human Population: Its Influences on Biological Diversity*, Ecological Studies 214,
DOI 10.1007/978-3-642-16707-2_6, © Springer-Verlag Berlin Heidelberg 2011

Table 6.1 Trends of urbanization by major areas

Percentage of population residing in urban areas				Projected annual rate of urbanization (%)
	1950	2000	2030	2005–2030
Africa	14.7	36.2	50.7	1.12
Asia	16.8	37.1	54.1	1.23
Europe	50.5	71.1	78.3	0.33
Latin America and the Caribbean	42.0	75.4	84.3	0.34
North America	63.9	79.1	86.7	0.29
Oceania	62.0	70.5	73.8	0.17
World	*29.0*	*46.7*	*59.9*	*0.83*

Source: United Nations, Department of Economic and Social Affairs, Population Division 2006. Used with permission.

humans, and direct management and its indirect effects constitute major forces shaping diversity in cities. Clearly, the assembly rules driving biological community structure in cities are very different from those driving *natural*, less human-dominated communities.

Biodiversity has different meanings to biologists, to policymakers, and to the public. Biologically, the term applies to many levels of biological organization (Noss 1990). Components of biodiversity include genetic diversity, species richness, and landscape diversity. Conservation biology often emphasizes the number of rare or endemic species within a community rather than simply focusing on the number of species of an area. However, most urban biodiversity studies focus on species richness indexed by the number of species in a given area, which is commonly called alpha-diversity (Magurran 1988). Another important metric of diversity, species evenness, is often reported as well. The focus of the following discussion is on species richness and species composition on the scale of habitats within urban areas.

6.2 Some Key Characteristics of Urban Ecosystems

As perhaps the most human-dominated system on the planet, cities represent a setting in which the effects of human demography on biodiversity may be most evident. Several human demographic trends are known to contribute to the impacts of urban areas on biodiversity. Increased urbanization – that is, increased population size of urban areas, which on average equals about 1.5 times the US national level of population growth – is due to both increases in the resident urban population and immigration from rural areas and abroad (Dow 2000; Cincotta et al. 2003, p. 53). Moreover, the area of most cities is expanding faster than their populations, a phenomenon known as urban sprawl (e.g., Alberti et al. 2003; Radeloff et al. 2005). This is due in part to shrinking household sizes (Liu et al. 2003) but also to larger parcel sizes in newer suburbs compared to older suburbs or central cities (Heimlich and Anderson 2001). The resulting conversion of wild or rural lands to urban lands generally produces reduced diversity of native flora and fauna and elevated numbers of exotic species (Kowarik 1995; Marzluff 2001; McKinney 2002), but there are exceptions to this pattern (Davis 1999; Samu and Szinetár 2000; Niemeleä et al. 2002). Furthermore, the kinds of human effects on biotic communities are far more complex than broad elimination of populations of native species or native habitat. Humans actively create biological communities in their parks, gardens, institutional grounds, and yards. The characteristics of these constructed communities depend on choices made by organizations, communities of people, households, and individuals (Odum 1970; Whitney and Adams 1980; Hope et al. 2003; Martin et al. 2004; Kinzig et al. 2005; Grove et al. 2006b). These agents and their decisions are in turn embedded in cultural traditions and socioeconomic networks (Machlis et al. 1997; Pickett et al. 1997). Thus, to understand the human drivers of the patterns of biodiversity in urban areas, we must find ways to integrate social science approaches with conventional ecological approaches to understanding biological communities (Cadenasso et al. 2006).

As humans actively construct biological communities in cities, they may juxtapose species that evolved on different continents and under different biophysical conditions (Hobbs et al. 2006). These novel communities often simultaneously have more *total* species but fewer *native* species than the surrounding native habitat (e.g., Marzluff 2001; McKinney 2002). The food resources provided by this novel habitat may actually be enriched relative to nonurban habitat, especially in temperate regions. For example, there may be a greater numbers of fleshy-fruited plants in urban sites (Beissinger and Osborne 1982). These drastic rearrangements of flora and fauna are thought by some to be leading toward a global homogenization of biotic communities and consequently a total reduction in global biological diversity (McKinney and Lockwood 1999; Blair 2001; Pouyat et al. 2006; Schwartz et al 2006). Regardless of whether this is true, the novel habitats created by humans clearly shape urban patterns of biodiversity. Urban flora and fauna are different from those of the surrounding areas.

Several factors account for the differences between urban and nonurban species assemblages. The modified urban environment may be suitable only to a subset of the original flora or fauna. This mainly depends on the ecological requirements of the species in question. However, while urban environment can be stressful for some species due to pollution, habitat fragmentation, etc., others may thrive in the cities because humans create favorable microhabitats or abundant resources for them. An example is the increased number of vine species present in forest canopy gaps in Baltimore, MD as compared to the smaller roster in rural forest gaps (Thompson 1999). These species are often nonnatives that have been associated with settlements and human activities for a long time, and which have been widely dispersed by people, or which readily spread on their own.

Other modified habitats also occur in urban areas in large numbers and include greenhouses, basements, compost piles, and green roofs. Many species have been described and are still known only from greenhouses (Korsós et al. 2002). Others, although first described from greenhouses, later spread into outdoor environments. One example is the common pillbug, *Armadillidium vulgare* Latr., in North America.

Using a combination of historical data and molecular approaches, Garthwaite et al. (1995) showed that *A. vulgare* on the east coast of the US was first reported from greenhouses but then subsequently was found in the south and west. Moreover, the west coast populations of *A. vulgare* are more similar to the Mediterranean populations in Europe, whereas the east coast populations are more similar to Atlantic European populations. This difference indicates independent introductions by different cultures on the two coasts, a pattern that corresponds well to the human immigration history of North America.

Urban environmental change is rapid and was until recently considered too fast for animals and plants to adapt. However, genetic evolution is a documented result of urbanization. On the Pacific coast of the United States, evolved changes in tail color were found in a population of birds in San Diego, CA (Yeh 2004). In plants, heavy metal tolerance in urban microhabitats or *brownfields* is an example of rapid evolution (e.g., Velguth and White 1998). Some species can also adapt nongenetically, that is, behaviorally or culturally, to the changes wrought by urbanization (Boyd and Richerson 1985; Yeh and Price 2004; Parker and Nilon 2008). For example, some birds have been documented to alter their songs in response to urban noise (Slabbekoorn and Peet 2003; Warren et al. 2006). Although these examples of populations adapting to the changing urban conditions join those of genetic change following industrialization (e.g., the textbook case of the peppered moth, *Biston betularia*), some species might disappear because they cannot adapt rapidly enough (genetically or behaviorally) to the novel conditions in an urban setting (Slabbekoorn and Peet 2003; Shochat et al. 2006; Warren et al. 2006).

6.3 Why Study Urban Biodiversity?

6.3.1 Values of Biodiversity

The reasons to study biodiversity in urban areas are many. Perhaps the most obvious reason is an esthetic or ethical one. Humans are attracted to nature and its living creatures. E.O. Wilson called this phenomenon “biophilia” and defined it as our “innate tendency to affiliate with life and lifelike processes” (Wilson 1984). Being surrounded by plants and animals creates a sense of peace and tranquility (Coley et al. 1997; Frumkin 2001). Given a choice, in the city people may prefer to live near a park, or have a view to a lake or river, assuming the social context and perceptions of hazards to be equivalent near such amenities and elsewhere. We surround ourselves with plants in our apartments or balconies (Rapoport 1993). We keep pets; we find having an aquarium in the living room relaxing. We plant shrubs and trees even in the tiniest yards and welcome birds with feeders in the winter. It does not matter whether these species have an ecological function (most of the time they do) or *just* esthetic value. If the birds help control harmful insects throughout spring and summer, that is an extra benefit; we just like having them around.

We liked caring for our pets even before scientific studies demonstrated that they lower blood pressure.

In spite of the profound esthetic values of urban biota, there are also many practical values or *ecosystem services* that are provided by urban biodiversity. Trees cool local climate, and together with herbaceous vegetation they take up excess nutrients and reduce runoff, but most people are unaware of these facts (McPherson et al. 1997). Community gardens utilize vacant spaces in a unique way. In addition to growing vegetables or flowers, they provide a meeting place for the neighborhood and promote social interaction (Burch and Grove 1993). They add to the city's green spaces and, by attracting pollinators and nectar feeders, they help maintain biodiversity (McIntyre and Hostetler 2001). Green roofs reduce storm water runoff, regulate building temperatures – thus conserving energy, and increase wildlife habitat area (Oberndorfer et al. 2007).

6.3.2 Roles of Exotic Species

In spite of the variety of values associated with urban vegetation, exotic species are often excluded from such recognition. The simplistic view that "exotics are always bad" needs revision. Exotics can be valuable in several ways. They may serve as important resources for native species. For example, in Davis, CA, 29 of 32 native butterflies breed on nonnative plants, many commonly designated as *weeds* (Shapiro 2002). Many of the native host plants no longer occur in the region, and exotics have taken on some of their important ecological roles. For example, various species of tamarisk (*Tamarix* spp.) have become important nesting sites for an endangered bird species. Many exotic species have existed in their nonnative host ecosystems for centuries and have integrated into the ecosystems to the extent that it is almost impossible to determine what the system must have been like before their arrival (Sukopp et al. 1990). Indeed, to return the biotic components of urban ecosystems to some ideal, pristine condition would be quite impossible in almost all cases of long residency.

A very special subset of exotic urban biodiversity is represented by species inhabiting artificial, human-made environments such as greenhouses and botanical gardens. As a result of the plant trade and the equable environment, the soil invertebrate species composition of some greenhouses is beyond imagination. For instance, a survey of only a few greenhouses in Hungary resulted in soil invertebrates originating from Asia, South America, and Africa (Csuzdi et al. 2007; Vilisics and Hornung 2009; Table 6.2). In general, the presence of such introduced species assemblages in greenhouses and similar novel environments are considered by ecologists to be undesirable, because they increase the chances for wider exotic introductions for which the ultimate consequences to native species and communities are yet unknown.

Taxonomists note that some species are only known from greenhouses (e.g., Korsós et al. 2002) or other human-engineered environments. The region of their

Table 6.2 Soil invertebrates and their known or possible zoogeographical origin from three greenhouses in Hungary

	Origin	Percentage of fauna of Hungary
Earthworms		
Phitemera bicincta (Perrier, 1875)	SE Asia	7.8
Amynthas corticis (Kinberg, 1867)	SE Asia	
Eudrilus eugeniae (Kinberg, 1867)	West Africa	
Ocnerodrilus occidentalis (Eisen, 1867)	Central America?	
Dendrobaena attemsi (Michaelsen, 1902)	Alps	
Microscolex phosphoreus (Dugés, 1837)	S America?	
Dichogaster bolaui (Michaelsen, 1891)	E Africa	
Isopods		
Armadillidium nasatum (Budde-Lund, 1885)	Atlanto-Mediterranean	7.3
Cordioniscus stebbingi (Patience, 1907)	Iberian	
Reductoniscus costulatus (Kesselyák, 1930)	SE Asia	
Trichorhina tomentosa (Budde-Lund, 1893)	Tropical America	
Millipedes		
Choneiulus palmatus (Nemec, 1895)		5
Cylindroiulus truncorum (Silvestri, 1896)	N. Africa	
Amphitomeus attemsi (Schubart, 1934)	S America	
Poratia digitata (Porat, 1889)	S America	
Cynedesmus formicola (Cook, 1896)	Canary Islands	

biogeographical origin is not known. Indeed, these species may no longer exist elsewhere. A few years ago a small centipede, *Nannarrup hoffmanni*, made headlines in the New York Times after being identified in the leaf litter in New York City's Central Park. It turned out to be a unique, newly described species whose discovery generated great excitement among taxonomists (Foddai et al. 2003). At present, Central Park is its only known locality. Soil and leaf litter generally harbor diverse communities, with many species undoubtedly awaiting description.

6.3.3 Contribution of Urban Studies to General Scientific Inquiry

The altered conditions and relatively fast rate of change in cities provide the basis for *natural experiments*, with rural or wild environments serving as controls, or more properly, reference systems. This situation provides urban ecologists with opportunities to observe and compare phenomena at the organismal, population, community, and ecosystem realms, and test general ecological hypotheses. An example is the relationship between disturbance and species diversity, which today goes beyond testing the intermediate disturbance hypothesis (Connell 1978). Going beyond intermediate disturbance requires identifying mechanisms by which potential invaders respond to specific human actions (Bart and Hartman 2000). This refinement examines specific events and responses rather than treating disturbance as a highly aggregated and hence inconsistent collection of diverse events.

In addition to serving as convenient experimental substrates, urban areas can serve as models of global change (Carreiro and Tripler 2005). The physical environment in cities, which includes elevated CO_2 concentrations, higher temperatures, and altered hydrological cycles, mimics key components of global climate change, thus providing opportunities to study responses of biota to such changes in existing rather than simulated environments. One such study in the Baltimore Metropolitan Region is underway (Ziska et al. 2004). Under these altered climatic conditions, shifts in species composition from natives to nonnatives, and from specialists to generalists, help to understand how redundancy may function in biological systems elsewhere as climate changes.

6.3.4 *Urban Ecosystems and Biodiversity Education*

Urban areas can serve as important venues for ecological and environmental education (Berkowitz et al. 2003). For most people, the only real encounter with the diversity of life happens in their city backyard or their suburban neighborhood. We have an obligation to help children and adults to learn about the species surrounding them and the role they play in that ecosystem (Miller 2006).

An example of engaging the public while gathering useful information on urban biodiversity is BioBlitz. The idea behind BioBlitz (http://www.pwrc.usgs.gov/blitz/) is to bring together taxonomists, park managers and the public as volunteer individuals, families, or school groups to document the biodiversity in their immediate environment. BioBlitz in cities usually takes place in urban parks and is a concentrated effort for a short period of time. Over a 24-h period, organisms are collected or observed and recorded for as many taxonomical groups as possible. It is not intended to be a rigorous scientific biodiversity survey; rather, it can serve as a starting point for scientific assessments. Since 1996, over 100 BioBlitz sites have been established all over the world. In addition to documenting species present in an area, the involvement of the public in such a fun event is a true benefit of this activity. People go home with a greater knowledge and appreciation of their own surroundings.

In other ongoing citizen science programs, instructions are given on websites along with simple taxonomic keys. Alternatively, schools or citizens are asked to send in specimens, and the resulting data are entered into a central database. The Wormwatch Program in Canada (http://www.naturewatch.ca/english/wormwatch/) and Walking with Woodlice in the UK (http://www.nhm.ac.uk/woodlice/) are examples. Survival and reproductive success of common backyard birds are the focus of the Nestwatch Program in the United States. (Marra and Reitsma 2001; Evans et al. 2004). DC Birdscape, though a coordinated efforts of several agencies and volunteers, systematically counts the birds in the Washington, DC area (Hadidian et al. 1997). Awareness and appreciation of urban biodiversity enhances the quality of life facilitating conservation efforts outside of the city as well (Savard et al. 2000) (Figs. 6.1 and 6.2).

Fig. 6.1 In the absence of large urban green spaces constructed plant assemblages, such green roofs contribute to human well-being, as well as maintaining higher biodiversity. (Photo was taken in Munich, Germany by K Szlavecz)

Fig. 6.2 Students sampling soil invertebrates in an urban forest in Baltimore, MD. Active participation in biodiversity surveys increases awareness and appreciation for the variety of life. (Photo was taken in Baltimore by K Szlavecz)

6.4 Research Approaches to Urban Biodiversity

Research on urban biodiversity must account for the spatial heterogeneity and complexity of urban ecosystems. The urban landscape is highly heterogeneous and exhibits striking changes from the rural surrounding to the urban core. Patchiness and the urban–rural gradient have been the two major factors guiding biodiversity assessments. These two perspectives account for the spatially complex nature of urban systems (Figs. 6.3 and 6.4).

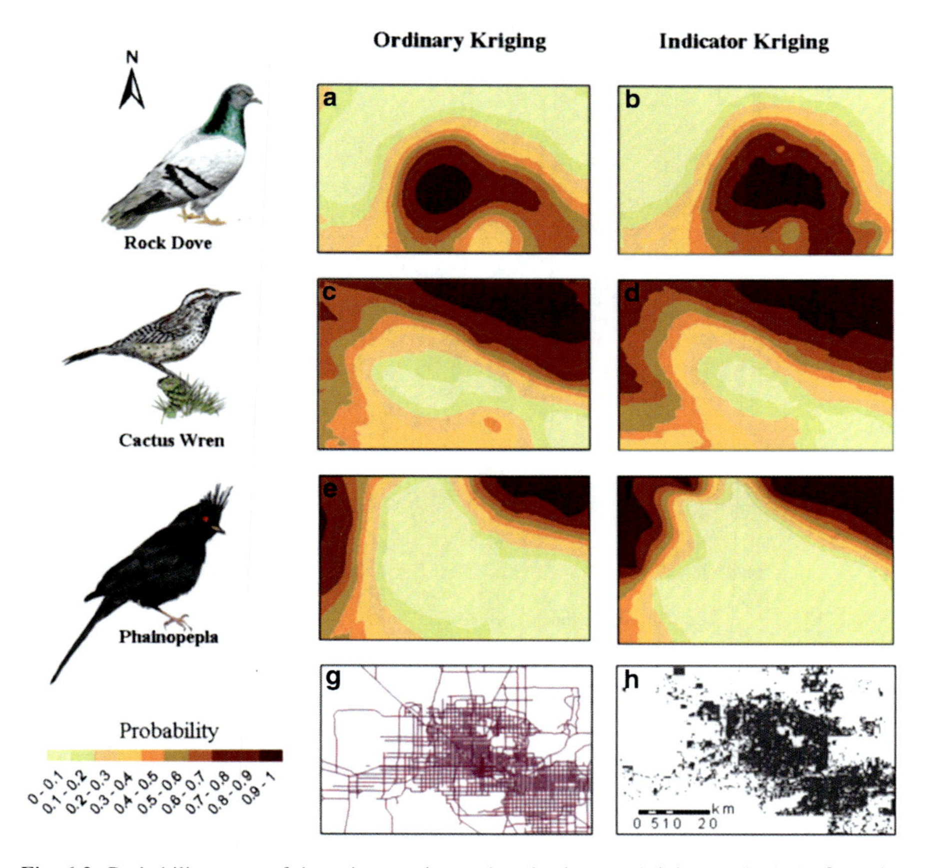

Fig. 6.3 Probability maps of the avian species analyzed using two kriging methods (**a–f**), major road (>3 lanes) networks (**g**) and urban land use (**h**) as defined by the Maricopa Association of Governments (2000). The stated probability indicates the likelihood that the number of individuals of a species will exceed the observed median. Three ecologically unique species are presented. Rock doves represent a species that is strongly linked to the urban ecosystem. The other two are native species. *Phainopepla* is an abundant species that very rarely utilizes the urban ecosystem. Whereas, the cactus wren encroaches into the urban core, it is more prevalent in the desert. Ordinary kriging was conducted with transformed data in order to satisfy the assumptions of normality. Indicator kriging was conducted with untransformed data. Images of birds are courtesy of Global Institute for Sustainability, Tempe, AZ. (Figure reproduced with permission from Walker et al. 2008)

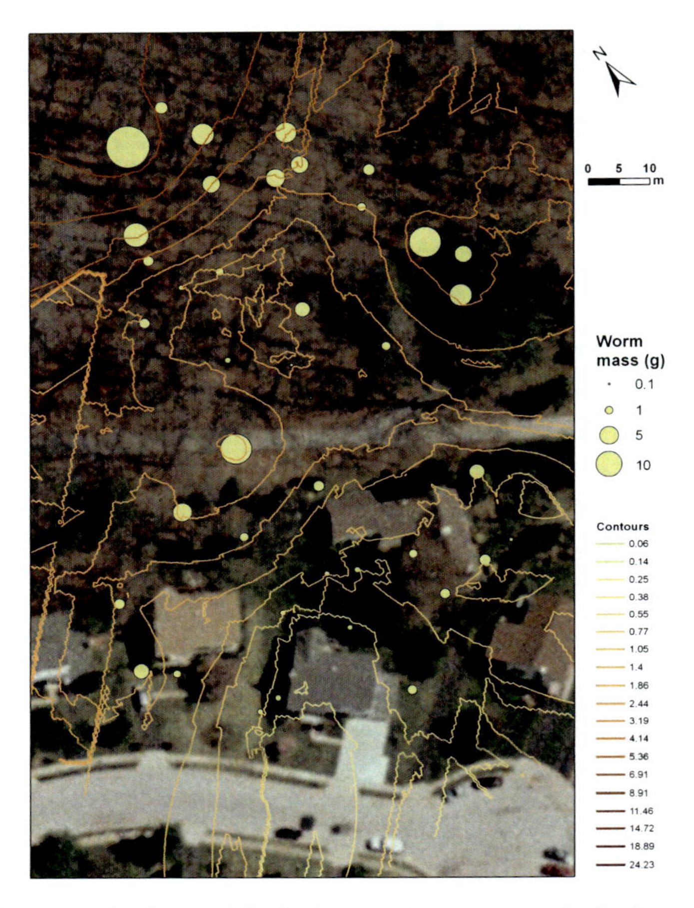

Fig. 6.4 Earthworm abundance and distribution on the heterogeneous urban landscape. The width of dots and contour lines indicate the number of earthworms collected from $25 \times 25\ \text{cm}^2$ quadrats. Data were collected in a typical suburban neighborhood, at the northern edge of Baltimore, Maryland. Data were compiled and the map was constructed by Erle Ellis, University of Maryland, Baltimore County

6.4.1 Urban–Rural Gradients

The urban–rural gradient approach is based on the assumption that key characteristics of the contrast between dense urban and the exurban fringe can be ordered conceptually. In rare cases, the ordering may be literal and follows a transect from the central business district to the rural fringe. In most cases, however, the complex mosaic of urban, suburban, exurban, and rural conditions are ordered in an abstract, multivariate space. Both of these approaches are captured within the concept of the *urban–rural gradient* (McDonnell et al. 1993). However, it is important to recognize that the conceptual gradient may not reflect any real transect on the ground in a specific metropolitan area. This is due to the fact that contemporary urban mosaics are multicentered, spatially extensive, and highly networked by transportation and information infrastructure (Bruegman 2005). The traditional concept of a single urban core with gradual transitions to the resource-supplying hinterlands is not suitable to the twenty-first century urban situation in either industrial or developing nations (Garreau 1991). Transects and abstract gradients have been powerful tools for studying both the biophysical and social variation in urban structure and function (Cadenasso et al. 2006; Dow 2000).

6.4.2 *Island Biogeography in Cities*

An important approach to the study of urban biodiversity has been to apply island biogeography theory (MacArthur and Wilson 1963, 1967) to cities. Island biogeography was, of course, developed for oceanic islands where isolation, local extinction, and the supply of propagules from continental source areas were all keys. This theory was applied to continental situations, such as mountain tops and lakes, early in its history. Ultimately, it was also applied to green space islands in cities. Such green islands result from land clearing associated with urban development and suburban sprawl, which leads to fragmentation of the existing native habitats. These isolated patches, surrounded by a different type of land use can be considered as habitat islands and, hence, subject to island biogeography theory. Urban biodiversity studies often use the theory of island biogeography to test hypotheses about patterns of species richness in these remnant habitats.

Of course, urban habitat fragments are different from oceanic islands in many ways. Urban habitats are primarily nonequilibrium systems with altered regimes of biophysical disturbance agents and continuous anthropogenic influence. Fire suppression and flood control are two common examples of altered disturbance regimes. Humans may also actively manage or use these habitat fragments, for example, by cutting and removing vegetation and trapping unwanted mammals. Recreation or collecting flowers, nuts, and mushrooms are examples of activities in urban habitat islands, which may affect their biodiversity. Colonization and extinction in the habitat fragments are influenced by humans, as they deliberately or accidentally transport species among these patches or they introduce nonnative species into the surrounding region. Alteration of the habitat may encourage more generalist species leading to altered species interactions, modified community structure, and ultimately altered ecosystem function. In light of such a suite of direct and indirect influencing factors, it is not surprising that urban species-area relationships are typically more complex than the textbook examples of island biogeography.

Numerous studies tested species-area and species-distance predictions in urban areas. Higher species richness in larger fragments was reported by Frundt and Ruszkowszki (1989), Sasvari (1984), Yamaguchi (2004), and Watson et al. (2005). Kitahara and Fuji (1997) successfully applied the theory of island biogeographical approach when analyzing the butterfly communities in newly designed parks in Tsukuba City, Japan. Schaefer and Kock (1979) found that ground beetle (Carabidae) species richness correlated with structural diversity of green islands better than with the island area in Kiel, Germany. Plant species composition and vertical structure were the determining factors for bird species richness in similar sized urban woodlots (Tilghman 1987), while landscape matrix was an additional factor in Australian bird assemblage structure (Watson et al. 2005). While distance of forest fragments from the urban core correlated well with carrion beetle (Silphidae) richness in Baltimore, the area of the fragments did not (Wolf and Gibbs 2004). Arthropod species richness did not vary with fragments size in Sydney, Australia (Gibb and Hochuli 2002), but species composition did. Lack of independence of the above multiple factors may lead to inconclusive results (Whitmore et al. 2002).

6.4.3 Patch Dynamics

The theory of patch dynamics incorporates refinements beyond the assumptions of the classical theory of island biogeography and, hence, has an important role to play in stimulating and organizing urban biodiversity studies (Pickett and Rogers 1997). Patch dynamics relaxes the assumption that areas between islands, or more broadly, any patch of interest, is completely hostile to the survival or movement of potential colonists to the focal patch. The area outside the focal patch can be considered to be differentially suitable for the survival, movement, and activities of organisms that also live in that patch. Thus, the matrix is not a uniformly homogeneous and hostile soup but is itself a heterogeneous patchwork or field of gradients that differentially affect organism performance and ecosystem function. As mentioned earlier, urban areas are manifestly patchy across space due to social differentiation, economic investment, a hierarchy of transportation networks, contrasting built structures, soils, substrates, impervious surfaces, and the presence and activity of different plants, fauna, and microbes (Cadenasso et al. 2006) (Figs. 6.1 and 6.2). Therefore, studying the extensive urban ecosystem as though it were a mosaic of contrasting patches or a field of gradients is a powerful way to address complexity in a realistic way. Indeed, new urban classifications are aimed at quantifying and portraying this patchiness in ways that integrate both social and biophysical structures (Cadenasso et al. 2007).

The changes in urban systems are also captured in the theory of patch dynamics. Urban areas grow on their edges, as illustrated by earlier citation of data on the disproportionate amount of land that is being converted to urban uses compared to the growth of urban population. In many postindustrial cities, internal dynamics are also conspicuous, as old industrial sites, underutilized shopping districts, and old residential areas are thinned or entirely abandoned. Disinvestment and depopulation defines an internal "urban frontier" that is equally dynamic if less extensive (Burch and Grove 1993).

A common theme of the studies of the structure of urban biodiversity relative to urban structure, demography, and human decision making is the emphasis on biotic species composition over species richness alone. Knowledge of species identity in an assemblage is important, because it allows further community analysis, such as detection of guild structure and the ratio of generalists to specialists or natives to nonnatives. Furthermore, the focus on species composition pinpoints fragments with unique species assemblages, which function as areas of exceptional biogeographical value. All of these biotic details help to make better conservation decisions (Roy et al. 1999; Gibb and Hochuli 2002; Watts and Lariviere 2004; Samu and Szinetár 2000; Schwartz et al. 2002; Smith et al. 2006a, b).

6.4.4 The Species Approach: Exotics and Introductions

Species introduction in urban systems is common, and species-based research is an important approach to urban systems. Urban and suburban areas often have higher species richness than surrounding wildlands. However, this is due to the addition of nonnative species in urban sites (Pyšek 1993, Kowarik 1995). Some of these species, such as ornamental plants, are deliberately brought to the cities, while others are introduced accidentally. This second class is exemplified by many soil organisms brought over as stowaways in ship ballasts (Lindroth 1957). Exotic species introduction and invasion has become a global environmental and economic issue (Pimentel et al. 2000). Accidentally introduced species are often synanthropic – species that tend to be associated with humans or their activities. These species are often generalists, and they typically move along with human settlement. Moreover, cities are considered to be highly disturbed environments, which tend to favor colonization of *weedy* species.

The similarity of urban environments and the subset of synanthropic species that follows human movement results in similar flora and fauna in worldwide, a phenomenon called biotic homogenization (McKinney 2006; McKinney and Lockwood 2001). This ecological process is of a great concern in conservation biology, because urban environments can serve as jumping-off points for exotics that can subsequently colonize rural or wild areas, often outcompeting native species. In addition, exotics can become nuisance or pest species as rats, cockroaches, fire ants, and the Asian longhorn beetle. It is beyond the scope of this chapter to analyze the factors that make good colonizers, or which contribute to susceptibility of potentially invasible environments to the establishment of a new invader. However, urban environments have much to contribute to understanding these important questions, which should be incorporated into urban ecological research. Biotic homogenization is of concern because the local and regional pools of biodiversity represent genetic capital which can be spent in future bouts of evolutionary adaptation to changing environments.

6.4.5 Human–Natural Coupling in Urban Systems

A further important approach to urban ecosystems is interdisciplinary study that incorporates both human drivers of biotic change and the response of humans to ecological changes (Cadenasso et al. 2006; Pickett et al. 2008). Integrating the full range of possible human drivers to studies of urban biodiversity requires extensive collaboration and interaction between social and natural scientists (Kinzig et al. 2000; Redman et al. 2004). Machlis et al. (1997) developed a Human Ecosystem Framework to guide researchers in selecting variables to consider in their models. They advocate using multiple variables to describe each parameter. Forester and Machlis (1996) applied an early version of this approach to understanding biodiversity loss at the national level using existing datasets. There are many datasets available for identifying social characteristics of subunits of urban areas such as neighborhoods or other patches. These include US Census data or similar datasets for other countries (Hope et al. 2003; Melles 2005), marketing analyses such as Claritas (Weiss 2000; Grove et al. 2006a), and municipal datasets (Grove et al. 2006b). These examples offer a starting place for considering the role of demography and human institutions in urban ecosystems.

An example of how demographic and social data can be linked with biodiversity and ecosystem performance data appears in Troy et al. (2007). That study analyzed the presence of woody and grassy vegetation and expenditure on their management in terms of the Claritas market-segmentation data. The authors discovered that the potential to add new vegetation on residential parcels in the city of Baltimore, MD, which constitute 60% of the land surface, was best predicted by the market segmentation based on 15 categories. This level of aggregation includes basic demographic variables but also includes aspects of social stratification. The complete market segmentation, recognizing 62 lifestyle clusters, best predicts variation in the actual amount of woody or grassy vegetation planted on residential parcels. Specific variables related to the realized amount of vegetation include family size, presence of detached homes, and marriage rates. Both the potential and the realized vegetation cover were related to population density, housing vacancy, and housing age. The percentage of African–American residents related negatively to expenditures on yard-related expenses, though the causes are not yet known.

For a richer understanding of the role that human values and actions play in patterns of biodiversity, we will need to forge collaborations between social scientists and natural scientists. This is happening in many places, including at two urban Long-term Ecological Research (LTER) sites were funded by the National Science Foundation: the Baltimore Ecosystem Study (BES LTER) and Central Arizona-Phoenix (CAP LTER) (Pickett et al. 1997; Grimm et al. 2000; Cadenasso et al. 2006; Pickett et al. 2008). These two LTER studies and other LTER sites that incorporate human demography and activities in their regional scope are exploring new territory for ecology. Researchers in such projects examine biodiversity and the ecological function of biotic assemblages throughout extensive mosaics. Thus, they examine the ecology *of* the city, rather than focusing only on the obvious green spaces. The more narrower approach can be labeled ecology *in* the city (Grimm et al. 2000). They examine the structure and dynamics of all sorts of patches in the urban mosaic. In addition, research teams jointly examine social and biophysical aspects of patch structure and function. Finally, partnerships with managers and policy makers ensure the relevance of surprisingly large proportions of such integrated research (Pickett et al. 2007).

One important component of how biodiversity functions in urban systems is especially lacking. Human values of biodiversity are little understood even in more natural environments. Some authors suggest that there are particular kinds of landscape preferences common to all humans (Orians and Heerwagen 1992; Ulrich 1993). But there is also variation in human preferences. One recent study used mail surveys to measure variation in human values of birds and their participation in activities that either benefit (e.g., feeding) or harm (e.g., owning free-roaming cats) birds along an urban to rural gradient (Lepczyk et al. 2004). Clergeau et al. (2001) also explored variation human perceptions of birds along an urban–rural gradient in France. Since the diversity of human groups in close proximity to one another reaches its peak in urban settings, cities represent a natural laboratory for exploring variation in human values for particular kinds of organisms and biodiversity. Results to date show surprising similarities in the environmental perceptions and values of wealthy and disadvantaged residents of some urban areas (Vemuri et al 2009), while at the same time it is clear that environmental hazards and amenities are disproportionately visited on poor or minority residents compared to wealthy and majority persons (Boone 2002).

6.4.6 Ecosystem Function

Ecosystem or habitat function is an additional important approach to understanding urban systems. Because, as mentioned before, urban areas contain many novel habitats and unprecedented environmental conditions, understanding the function of these habitats becomes all the more critical. More studies need to incorporate a mechanistic approach to understanding urban biotic communities rather than simply documenting the presence and number of various species. When mechanistic studies have been undertaken, there are sometimes conflicting results. For example, some studies suggest that levels of predation on birds are lower in cities (reviewed in Shochat 2004), but others suggest that predators occur at even higher densities in the city than in the surrounding countryside (Sorace 2002). Studies of soil invertebrates have addressed mechanistic questions such as the effect of species composition on rates of decomposition (Broll and Keplin 1995) and biogeochemical cycling (Steinberg et al. 1997; Pavao-Zuckerman and Coleman 2005; Szlavecz et al. 2006), but there are far fewer studies of invertebrates than of the more charismatic species such as birds.

The ecosystem function of different species, especially exotics vs. natives, is a significant component of the functional urban approach. A specific example of such research is a study by Groffman and Crawford (2003) in Baltimore. Sites dominated by the exotic *Ailanthus altissima* and *Acer platanoides* had higher levels of soil nitrate and nitrification rates than sites dominated by native trees. Thus, exotic species accelerate both nitrogen availability and loss. Sites dominated by the exotics trees also had higher soil moisture. It is not known whether the exotics colonize wetter sites or alter water balance in sites they invade.

Ehrenfeld (2003), in a wide-ranging literature review of the functions of exotic species in urban systems, concluded that invasive exotic plants often increase biomass and net primary production. Furthermore, exotics increase the availability and change fixation of nitrogen, resulting in litter of higher quality that tends to decompose faster than litter of native species. However, not all exotic species produce these results, with the opposite relationships sometimes found. There are no patterns in other components of nutrient dynamics such as soil pools of carbon and nitrogen. The same species can behave differently at different sites, suggesting that specific environmental conditions may determine the nature of ecosystem effects. Exotic plants differ from native species in terms of biomass, rates of productivity, tissue chemistry, plant morphology, and seasonal timing.

6.4.7 Overcoming Logistic Problems

Whatever research approaches are used in urban systems, there are logistical problems. The use of experiments is one of these approaches that are problematic in urban settings. To move beyond correlative studies, ecologists prefer to use manipulative field experiments with proper replication and control. However, unless the experiments are small scale, such an approach poses a problem in cities, due to highly parcelized land ownership, zoning regulations, access, and other issues (Cook et al. 2004). Scientists have to work closely with the city government and its departments, the residents and various organizations, and all of this puts constraints on ecological experiments. The experimental method itself, as mentioned before, is sometimes a constraint. Manipulating densities of exotic species to measure their effect on native species or on ecosystem function can be impractical, unpleasant (e.g., involving pest species), or simply intrusive. In addition, in order to conduct research in urban habitats, scientists must engage residents (Pickett 2003). Involvement of residents is essential, since we often sample on their properties – and much taxonomic diversity lies in backyards, particularly among small species that have large influences on nutrient cycling (Wilson 1987; Grove 1995; Kim and Byrne 2006). For instance, to look at how lawn care might affect soil arthropod communities involves cooperation from the residents who are asked to fill out questionnaires and allow sampling on their lawn. Methods have to be modified in residential areas, because traditional ecological sampling techniques are sometimes destructive of the habitat (e.g., removal of large quantities of soil) or involve use of potentially harmful substances. Use of water instead of propylene glycol or ethanol as collecting fluid in pitfall traps (McIntyre et al. 2001) requires shorter sampling periods, or more frequent visits to the sampling sites. Proper assessment of these groups (arthropods, fungi, nematodes, etc) requires taxonomic knowledge or the availability of taxonomic services, both of which are declining (Kim and Byrne 2006).

In our experience, most citizens have been open and willing to help when they can provide valuable data for biodiversity monitoring studies. Such cooperative efforts are successful if good, long-term working relationships with homeowners are maintained, clear and simple instructions are provided about data collection, and when necessary, expert help is provided about their environmental concerns.

In spite of the difficulties of conducting experimental research in cities, there are successes. Cook et al. (2004) illustrate a compelling experiment on landscaping practices and management conducted with the cooperation of a large land owner and local residents. Felson and Pickett (2005) suggest that a closer partnership between scientists and urban designers can help solve this problem more generally. A close collaboration provides the opportunity for rigorous experimentation while creating esthetic and functional urban spaces at the same time. Such spaces are installed for the long term. Thus, the experiments will provide an invaluable tool for monitoring, and reveal urban ecological patterns and processes, including those related to biodiversity.

We must be ready to exploit *natural experiments* such as accidental introductions of new species or major urban development or revitalization projects (Pickett et al. 2007). A project in west Baltimore illustrates many of the characteristics of these experiments. This project combines neighborhood revitalization, greening, removal of unnecessary asphalt, and installing best-management practices for storm water control (Pickett et al. 2007). The interventions involve the government agencies, local communities, and can be used by researchers to improve general scientific knowledge. Importantly, the project also serves to inform local communities about their environment and the way it is changing and is useful in local schools as well.

6.5 Questions and Research Directions on Urban Biodiversity

Throughout the above discussion, we have identified questions and data gaps motivating future research on urban biodiversity. Because the field of urban ecological studies and its integration with social, economic, and physical sciences is evolving so rapidly, we highlight research opportunities in this section.

Variation in biodiversity within urban areas is often associated with such straightforward social and demographic factors as human population density, building density, and the amount of impervious surface (DeGraaf and Wentworth 1986; Munyenyembe et al. 1989; Blair 1996; Clergeau et al. 1998; Fernández-Juricic and Jokimaki 2001). Studies of urbanization gradients have successfully employed such measures to describe urban patterns of diversity for a variety of taxa (Blair 1999; Germaine and Wakeling 2001; Pickett et al. 2001; McKinney 2002). Even when combined to generate multivariate indices of urbanization, however, these studies often leave much of the variation in diversity across the urban matrix unexplained. This approach does not identify the full richness of social and economic forces driving changes in biodiversity over time.

New approaches are necessary in order to understand the complex human social, cultural, and economic forces underlying changing patterns of biodiversity. Standard human socioeconomic data may identify some of the human activities that influence biological communities (Grove and Burch 1997; Grove et al. 2002). In residential areas, household income may strongly influence the capacity for homeowners to construct diverse plant communities in their neighborhoods (Schmid 1975; Whitney and Adams 1980; Hope et al. 2003; Kinzig et al. 2005). Beyond access to resources, however, differences in lifestyle or ethnicity are associated with differences in landscape preferences, leading to different patterns of biodiversity (Kaplan and Talbot 1988; Fraser and Kenney 2000; Kinzig et al. 2005; Grove et al. 2006b). In Phoenix, AZ, the number of bird species in small urban parks is strongly correlated with the socioeconomic status of surrounding neighborhoods containing higher bird diversity in wealthier neighborhoods (Kinzig et al. 2005). In Baltimore, MD, the nature and cover of vegetation was determined by population density, aggregated neighborhood wealth, or lifestyle indices, depending on whether vegetation in public rights of way, the neglected riparian commons, or private property was examined (Grove et al. 2006b).

Many research questions remain: What mechanisms mediate the relationship between human socioeconomic factors and patterns of species diversity in specific situations? To what extent are these patterns dependent on the aggregate effects of individual human behaviors such as feeding birds or gardening vs. government level decision making such as zoning or redevelopment plans (Kinzig et al. 2005)? Does education about the environment alter decision-making that affects biodiversity such as landscaping decisions (Berkowitz et al. 2003)? What is the relationship of urban human population to the plants and animals that occupy nonurban areas? For most people in the inner city, animals mean pests, including cockroaches, rats, and ants, and perhaps less-than-glamorous species such as house sparrows and pigeons. What kinds of biological diversity do humans value? The answers to these questions are essential to predicting the future outcome of urban biodiversity and to motivating efforts to conserve biodiversity in urban areas – the setting in which the majority of humans now live and work.

The question of urban sustainability looms large as more of humanity finds itself in cities and complex urban agglomerations. In this context, one exciting question is the characteristics of organisms that will successfully maintain populations in a highly modified environment (Chace and Walsh 2006). Genetics, physiology, and behavior are key components (e.g., Harris and Trewhella 1988; Wandeler et al. 2003). However, it will also be necessary to monitor and understand the causes and consequences of biodiversity change in evolving urban areas, because the changes are often nonlinear, and may appear with considerable time-lag (Hansen et al. 2005). The traditional view that urbanization necessarily and uniformly leads to an impoverished flora and fauna must be revised, as the patterns appear to be more complex (Pickett et al. 2001, 2008; Schwartz et al. 2001, 2002). The opposite view that urban flora and fauna can be enriched due to the presence of many nonnatives is also simplistic (Marzluff and Rodewald 2008).

Biodiversity changes are not only due to direct human influences but also due to the interactions within the biotic community itself. Are the biological forces shaping community structure altered in cities? How did the disappearance from urban habitats of predators, such as wolves, foxes, or cougars, affect the dynamics of prey populations, such as squirrels, rabbits, or deer? How does reemergence of some of these species (Gompper 2002; Morey et al 2007) modify urban trophic structure? Are urban soil food webs highly modified, and do these differences along with altered abiotic conditions affect decomposition processes (Carreiro et al. 1999; Walton et al. 2006)?

What is the effect of human control on prey populations? Are there effective controls possible for emerging outbreaks such as that of white-tailed deer in the eastern US Megalopolis? What are the patterns of herbivory in plant assemblages characterized by introduced, ornamental plants?

The understanding of the urban biota that would emerge from broadly answering the questions posed so far can be considered a prolog to this question about the feedback between biota and the remaining components of urban ecosystems: Do altered patterns of urban biodiversity lead to altered ecosystem functions? For example, how do biomass and primary productivity of lawns compare to natural grasslands? How does kind and level of lawn maintenance affect this function? How do bird feeders change feeding patterns and survival of birds? Are ecosystem functions that yield ecosystem services in urban areas in fact sustainable? Answering this question will require ecologists to understand better the functioning of the novel habitats constructed in urban areas. Most ecological research in cities has taken place in patches that are analogs of the natural or wild habitats' ecologists who have been investigating outside cities for a hundred years. Under what conditions do natives out-compete exotics and vice versa? Do exotic species alter interactions among native species such as predators and their prey, parasites and their hosts, and plants with their pollinators? Many birds and some insects have been observed feeding on nonnative plant species (Reichard et al. 2001; Shapiro 2002). To what extent do these novel food sources and host plants provide usable habitat for native animal species? These questions remain largely unaddressed in the literature.

If humans around the world desire similar features from their environment, then urban areas may converge with one another in their species composition (McKinney and Lockwood 1999; McKinney 2002; Pouyat et al. 2006). This might happen both because humans actively transport the species between continents and because similar characteristics of urban environments select for similar kinds of species – that is, species that can adapt to the presence of humans, their built structures, and the high heterogeneity of urban environments. In addition, urban heat islands, soil moisture conditions, and calcium enrichment from concrete and imported limestone and marble contribute to environmental homogenization (Pouyat et al 2010). So far, there have been few empirical tests of the global homogenization or global convergence hypothesis (but see Blair 2001).

An alternative, not mutually exclusive, possibility is that urbanization could generate new biological diversity through evolutionary responses of species to the novel selection pressures found in cities (Slabbekoorn and Peet 2003). Likewise, different habitats found in different cities might generate divergent selection pressures, resulting in the evolution of new, genetically distinct populations from city to city. House finches residing in highly urbanized settings have shown rapid morphological evolution, forming genetically distinct populations (Badyaev et al. 2000). Tamarisk, an exotic, invasive species in the US, has evolved genetically distinct and novel variants via hybridization, variants that only occur in the US (Gaskin and Schaal 2002). Furthermore, this exotic apparently serves as nesting habitat for a native endangered bird species, the Southwest Willow Flycatcher (Ohmart et al. 1991; Sogge et al. 1997). Introduced populations of gray squirrels in Europe appear to differ behaviorally and perhaps genetically from their North American ancestors (Parker and Nilon 2008). Continuing biodiversity research in urban areas will surely reveal more cases of evolutionary responses to novel habitats created by human actions.

Recently Swan et al. (2010) proposed metacommunity theory as a useful tool to explain patterns of diversity in urban areas. This concept integrates local (environmental factors, species interactions) and regional (species pool, dispersal) factors as forces shaping plant and animal communities. By combining physical, biological, and socioeconomic processes at various scales, the urban metacommunity concept can generate testable hypotheses leading to a better understanding of urban biodiversity.

6.6 Conclusions

Urban biodiversity is at once both an old subject and one that is exploring new frontiers, encompassing the urban fringe and the sometimes rapidly changing patches within old urban cores. Urban areas have been shown to be surprisingly speciose habitats, with combinations of natives and exotics, generalist invaders adapted to human activities, and rapidly evolving specialists on novel and stressful habitats. Foundation studies of urban biota are beginning to be complemented by molecular genetic research, behavioral studies, landscape and dispersal research, and community interaction research. Increasingly, these disciplinary-based endeavors are being placed in an integrative framework that attempts to capture the dynamism of an urban matrix driven by complex feedbacks between ecological, social, economic, and physical processes. The question of human geography and its interaction with biodiversity is core to this rapidly expanding field of urban ecosystem research.

Acknowledgments The Baltimore Ecosystem Study is supported by the National Science Foundation Long-Term Ecological Research program, grant number DEB 0423476. We also acknowledge financial support from Hungarian Science Research Fund (OTKA No. T43508, OTKA T42745). We thank the editors, Richard Cincotta and Larry Gorenflo for inviting us to write a chapter, for their help in editing the manuscript, and most importantly for their patience during this process.

References

Alberti M, Marzluff JM, Shulenberger E, Bradley G, Ryan C, Zumbrunnen C (2003) Integrating humans into ecology: opportunities and challenges for studying urban ecosystems. BioScience 53:1169–1179
Badyaev A, Hill G, Stoehr A, Nolan P, Mcgraw K (2000) The evolution of sexual size dimorphism in the house finch. II. Population divergence in relation to local selection. Evolution 54:2134–44
Bart D, Hartman JM (2000) Environmental determinants of *Phragmites australis* expansion in a New Jersey salt marsh: an experimental approach. Oikos 81:59–69
Beissinger SR, Osborne DR (1982) Effects of urbanization on avian community organization. Condor 84:75–83
Berkowitz AR, Nilon C, Hollweg KS (eds) (2003) Understanding urban ecosystems: a new frontier for science and education. Springer, New York
Blair RB (1996) Land use and avian species diversity along an urban gradient. Ecol Appl 6:506–519
Blair RB (1999) Birds and butterflies: surrogate taxa for assessing biodiversity? Ecol Appl 9:164–170
Blair RB (2001) Creating a homogeous avifauna. In: Marzluff JM, Bowman R, Donnelly R (eds) Avian ecology in an urbanizing world. Kluwer, Boston, pp 459–486
Boone CG (2002) An assessment and explanation of environmental inequity in Baltimore. Urban Geogr 23:581–595
Botkin DB, Beveridge CE (1997) Cities as environments. Urban Ecosyst 1:3–19
Boyd R, Richerson PJ (1985) Culture and the evolutionary process. University of Chicago Press, Chicago
Brazel A, Selover N, Vose R, Heisler G (2000) The tale of two cities – Baltimore and Phoenix urban LTER sites. Climate Res 15:123–35
Broll G, Keplin B (1995) Ecological studies of urban lawns. In: Sukopp H, Numata M, Huber A (eds) Urban ecology as a basis for urban planning. SPB Academic, Amsterdam, pp 71–82
Bruegman R (2005) Sprawl: a compact history. University of Chicago Press, Chicago, IL
Burch WR Jr, Grove JM (1993) People, trees, and participation on the urban frontier. Unasylva 44:19–27
Cadenasso ML, Pickett ST, Grove JM (2006) Integrative approaches to investigating human-natural systems: the Baltimore Ecosystem Study. Nat Sci Soc 14:1–14
Cadenasso ML, Pickett STA, Schwarz K (2007) Spatial heterogeneity in urban ecosystems: reconceptualizing land cover and a framework for classification. Front Ecol Environ 5:80–88
Carreiro MM, Tripler CE (2005) Forest remnants along urban-rural gradients: examining their potential for global change research. Ecosystems 8:568–582
Carreiro MM, Howe K, Parkhurst DF, Pouyat RV (1999) Variation in quality and decomposability of red oak leaf litter along an urban-rural gradient. Biol Fertil Soils 30:258–268
Chace JF, Walsh JJ (2006) Urban effects on native avifauna: a review. Landsc Urban Plan 74:46–69
Cincotta RP, Engelman R, Anastasion D (2003) The security demographic: population and civil conflict after the Cold War. Population Action International, Washington, DC
Clergeau P, Savard JPL, Mennenchez G, Falardeau G (1998) Bird abundance and diversity along an urban-rural gradient: a comparative study between two cities on different continents. Condor 100:413–425
Clergeau P, Jokimaki J, Savard JPL (2001) Are urban bird communities influenced by the bird diversity of adjacent landscapes? J Appl Ecol 38:1122–1134
Coley RL, Sullivan WC, Fe K (1997) Where does community grow? The social context created by nature in urban public housing. Environ Behav 29:468–494
Connell JH (1978) Diversity in tropical rain forests and coral reefs. Science 199:1302–1310
Cook WM, Casagrande DG, Hope D, Groffman PM, Collins SL (2004) Learning to roll with the punches: adaptive experimentation in human-dominated systems. Front Ecol Env 2:467–474

Csuzdi CS, Pavlicek T, Nevo E (2007) Is *Dichogaster bolaui* (Michaelsen, 1891) the first domicole earthworm species? Eur J Soil Biol. doi:10.1016/j.ejsobi.2007.05.003 DOI:dx.doi.org

Davis CA (1999) Plant surveys and searches for rare vascular plant species at two pilot areas: Gwynns Falls/Leakin Park, Baltimore City, MD. Natural History Society of Maryland, Baltimore, pp 1–26

DeGraaf RM, Wentworth JM (1986) Avian guild structure and habitat associations in suburban bird communities. Urban Ecol 9:399–412

Dow K (2000) Social dimensions of gradients in urban ecosystems. Urban Ecosyst 4:255–275

Ehrenfeld JG (2003) Effects of exotic plant invasions on soil nutrient cycling Processes. Ecosystems 6:503–523

Evans E, Abrams E, Roux K, Salmonsen L, Reitsma R, Marra PP (2004) The neighborhood nestwatch program: sense of place and science literacy in a citizen-based ecological research project. Conserv Biol 19:589–584

Felson AG, Pickett STA (2005) Designed experiments: new approaches to studying urban ecosystems. Front Ecol Environ 3:549–556

Fernández-Juricic E, Jokimaki J (2001) A habitat island approach to conserving birds in urban landscapes: case studies from southern and northern Europe. Biodivers Conserv 10:2023–2043

Foddai D, Bonato L, Pereira LA, Minelli A (2003) Phylogeny and systematics of the Arrupinae (*Chilopoda: Geophilomorpha: Mecistocephalidae*) with the description of a new dwarfed species. J Nat Hist 37:1247–1267

Forester DJ, Machlis GE (1996) Modeling human factors that affect the loss of biodiversity. Conserv Biol 10:1253–1263

Fraser EDG, Kenney WA (2000) Cultural background and landscape history as factors affecting perceptions of the urban forest. J Arboric 26:106–112

Frumkin H (2001) Beyond toxicity: human health and the natural environment. Am J Prev Med 20:234–240

Frundt HC, Ruszkowszki B (1989) Untersuchungen zur biologie stadtischer boden. 4. Regenwurmer. Asseln und Diplopoden. Verh Ges f Okol 18:193–200

Garreau J (1991) Edge city: life on the new frontier. Doubleday, New York

Garthwaite RL, Lawson R, Sassaman C (1995) Population genetics of *Armadillidium vulgare* in Europe and North America. Crustacean Issues 9:145–199

Gaskin JF, Schaal BA (2002) Hybrid Tamarix widespread in US invasion and undetected in native Asian range. Proc Natl Acad Sci USA 99:11256–11259

Germaine SS, Wakeling BF (2001) Lizard species distributions and habitat occupation along an urban gradient in Tucson, Arizona, USA. Biol Conserv 97:229–237

Gibb H, Hochuli DF (2002) Habitat fragmentation in an urban environment: large and small fragments support different arthropod assemblages. Biol Conserv 106:91–100

Gompper ME (2002) Top carnivores in the suburbs? Ecological and conservation issues raised by colonization of north-eastern North America by coyotes. BioScience 52:185–190

Grimm NB, Grove JM, Redman CL, Pickett STA (2000) Integrated approaches to long-term studies of urban ecological systems. BioScience 50:571–584

Groffman PM, Crawford MK (2003) Denitrification potential in urban riparian zones. J Environ Qual 32:1144–1149

Groffman PM, Bain DJ, Band LE, Belt KT, Brush GS, Grove JM, Pouyat RV, Yesilonis IC, Zipperer WC (2003) Down by the riverside: urban riparian ecology. Front Ecol Environ 6: 315–321

Grove JM (1995) Excuse me, could I speak to the property owner please? Common Property Resour Digest 35:7–8

Grove JM, Burch WR (1997) A social ecology approach to urban ecosystem and landscape analyses. Urban Ecosyst 1:259–275

Grove JM, Schweik C, Evans T, Green G (2002) Modeling human-environmental systems. In: Clarke KC, Parks BE, Crane MP (eds) Geographic information systems and environmental modeling. Prentice-Hall, Book City, pp 160–188

Grove JM, Cadenasso ML, Burch WR Jr, Pickett STA, Schwarz K, O'Neill-Dunne JPM, Wilson MA, Troy A, Boone CG (2006a) Data and methods comparing social structure and vegetation structure of urban neighborhoods in Baltimore, Maryland. Soc Nat Resour 19:117–136

Grove JM, Troy AR, O'Neil-Dunne JPM, Burch WR, Cadenasso ML, Pickett STA (2006b) Characterization of households and its implications for the vegetation of urban ecosystems. Ecosystems 9:578–597

Hadidian J, Sauer J, Swarth C, Handly P, Droege S, Williams C, Huff J, Didden G (1997) A citywide breeding bird survey for Washington, DC. Urban Ecosyst 1:87–102

Hansen AJ, Knight RL, Marzluff JM, Powell S, Brown K, Gude P, Jones K (2005) Effects of exurban development on biodiversity: patterns, mechanisms and research needs. Ecol Appl 15:1893–1905

Harris S, Trewhella WJ (1988) An analysis of some of the factors affecting dispersal in an urban fox (*Vulpes vulpes*) population. J Appl Ecol 25:409–422

Heimlich RE, Anderson WD (2001) Development at the urban fringe and beyond: impacts on agriculture and rural land. US Department of Agriculture, Washington DC

Hobbs RJ, Arico S, Baron J, Bridgewater P, Cramer VA, Epstein PR, Ewel JJ, Klink CA, Lugo AE, Norton D, Ojima D, Richardson DM, Sanderson EW, Valladares F, Vila M, Zamora R, Zobel M (2006) Novel ecosystems: theoretical and management aspects of the new ecological world order. Global Ecol Biogeogr 15:1–7

Hope D, Gries C, Zhu W, Fagan WF, Redman CL, Grimm NB, Nelson A, Martin C, Kinzig A (2003) Socio-economics drive urban plant diversity. Proc Natl Acad Sci USA 100:8788–8792

Kaplan R, Talbot JF (1988) Ethnicity and preference for natural settings: a review and recent findings. Landsc Urban Plan 15(1–2):107–117

Kim KC, Byrne LB (2006) Biodiversity loss and the taxonomic bottleneck: emerging biodiversity science. Ecol Res 21:794–810

Kinzig AP, Carpenter S, Dove M, Heal G, Levin SA, Lubchenco J, Schneider S, Starrett D (2000) Nature and society: an imperative for integrated research. Report to the National Science Foundation. http://www.public.asu.edu/~akinzig/nsfes.pdf

Kinzig AP, Warren PS, Martin C, Hope D, Katti M (2005) The effects of human socioeconomic status and cultural characteristics on urban patterns of biodiversity. Ecol Soc 10:23

Kitahara M, Fuji K (1997) An island biogeographical approach to the analysis of butterfly community patterns in newly designed parks. Res Popul Ecol 39:23–35

Korsós Z, Hornung E, Kontschán J, Szlavecz K (2002) Isopoda and Diplopoda of urban habitats: new data to the fauna of Budapest. Ann Hist Nat Mus Nat Hung 94:193–208

Kowarik I (1995) On the role of alien species in urban flora and vegetation. In: Pyšek P (ed) Plant invasions: general aspects and special problems. SPB Academic, Amsterdam, pp 85–103

Lepczyk CA, Mertig AG, Liu J (2004) Assessing landowner activities related to birds across rural-to-urban landscapes. Environ Manage 33:110–125

Lindroth CH (1957) The faunal connections between Europe and North America. Wiley, New York, p 344

Liu J, Daily GC, Ehrlich PR, Luck GW (2003) Effects of household dynamics on resource consumption and biodiversity. Nature 421:530–533

MacArthur RH, Wilson EO (1963) The equilibrium theory of insular biogeography. Evolution 17:373–387

MacArthur RH, Wilson EO (1967) The theory of island biogeography. Princeton University Press, Princeton

Machlis GE, Force JE, Burch WR (1997) The human ecosystem. 1. The human ecosystem as an organizing concept in ecosystem management. Soc Nat Resour 10:347–367

Magurran AE (1988) Ecological diversity and its measurement. Princeton University Press, Princeton

Maricopa Association of Governments (2000) Urban land use. http://magwww.mag.maricopa.gov/detail.cms?item=2570

Marra PP, Reitsma R (2001) Neighborhood nestwatch: science in the city. Wild Earth (Fall/Winter):28–30

Martin CA, Warren PS, Kinzig AP (2004) Neighborhood socioeconomic status is a useful predictor of perennial landscape vegetation in residential neighborhoods and embedded small parks of Phoenix, AZ. Landsc Urban Plan 69:355–368

Marzluff JM (2001) Worldwide urbanization and its effects on birds. In: Marzluff JM, Bowman R, Donnelly R (eds) Avian ecology in an urbanizing world. Kluwer Academic, Boston, pp 19–47

Marzluff J, Rodewald A (2008) Conserving biodiversity in urbanizing areas: nontraditional views from a bird's perspective. Cities Environ 1:1–27. http://escholarship.bc.edu/cate/vol1/iss2/6

McDonnell MJ, Pickett STA, Pouyat RV (1993) The application of the ecological gradient paradigm to the study of urban effects. In: McDonnell MJ, Pickett STA (eds) Humans as components of ecosystems: the ecology of subtle human effects and populated areas. Springer, New York, pp 175–189

McIntyre NE, Rango J, Fagan WF, Faeth SH (2001) Ground arthropod community structure in a heterogeneous urban environment. Landsc Urban Plan 52:257–274

McKinney ML (2002) Urbanization, biodiversity and conservation. BioScience 52:883–890

McKinney ML (2006) Urbanization as a major cause of biotic homogenization. Biol Conserv 127:247–260

McKinney ML, Lockwood JL (1999) Biotic homogenization: a few winners replacing many losers in the next mass extinction. Trends Ecol Evol 14:450–453

McKinney ML, Lockwood JL (2001) Biotic homogenization: a sequential and selective process. In: Lockwood JL, McKinney ML (eds) Biotic homogenization. Kluwer Academic, New York, pp 1–17

Mclntyre NE, Hostetler ME (2001) Effects of urban land use on pollinator (*Hymenoptera: Apoidea*) communities in a desert metropolis. Basic Appl Ecol 2:209–218

McPherson EG, Nowak D, Heisler G, Grimmond S, Souch C, Grant R, Rowntree R (1997) Quantifying urban forest structure, function, and value: the Chicago Urban Forest Climate Project. Urban Ecosyst 1:49–61

Melles S (2005) Urban bird diversity as an indicator of human social diversity and economic inequality in Vancouver, British Columbia. Urban Habitats 3:25–48

Miller JR (2006) Biodiversity conservation and the extinction of experience. Trends Ecol Evol 20:430–434

Morey PS, Gese EM, Gehrt S (2007) Spatial and temporal variation in the diet of coyotes in the Chicago metropolitan area. Am Midl Nat 158:147–161

Munyenyembe F, Harris J, Hone J, Nix H (1989) Determinants of bird populations in an urban area. Aust J Ecol 14:549–557

Niemeleä J, Kotze DJ, Venn S, Penev L, Stoyanov I, Spence J, Hartley D, Montes de Oca E (2002) Carabid beetle assemblages (Coleoptera: Carabidae) across urban-rural gradients: an international comparison. Landsc Ecol 17:397–401

Noss RF (1990) Indicators for monitoring biodiverisity: a hierarchical approach. Conserv Biol 4:355–364

Oberndorfer E, Lundholm J, Bass B, Coffman RR, Doshi H, Dunett N, Gaffin S, Kohler M, Liu KKY, Rowe B (2007) Green roofs as urban ecosystems: ecological structures, functions, and services. BioScience 57:823–833

Odum HT (1970) Environment, power, and society. Wiley-Interscience, New York

Ohmart RD, Hunter WC, Rosenberg KV (1991) Birds of the lower Colorado River Valley. University of Arizona Press, Arizona

Orians GH, Heerwagen JH (1992) Evolved responses to landscapes. In: Narkow JH, Cosmides L, Tooby J (eds) The adapted mind: evolutionary psychology and the generation of culture. Oxford University Press, New York

Parker T, Nilon C (2008) Gray squirrel density, habitat suitability and behavior in urban parks. Urban Ecosyst 11:243–255

Paul MJ, Meyer JL (2001) Streams in the urban landscape. Annu Rev Ecol Syst 32:333–365

Pavao-Zuckerman MA, Coleman DC (2005) Decomposition of chestnut oak (*Quercus prinus*) leaves and nitrogen mineralization in an urban environment. Biol Fertil Soils 41:343–349

Pickett STA (2003) Why is developing a broad understanding of urban ecosystems important to science and scientists? In: Berkowitz AR, Nilon CH, Hollweg KS (eds) Understanding urban ecosystems: a new frontier for science and education. Springer, New York, pp 58–72

Pickett STA, Rogers KH (1997) Patch dynamics: the transformation of landscape structure and function. In: Bissonette JA (ed) Wildlife and landscape ecology: effects of pattern and scale. Springer, New York, pp 101–127

Pickett STA, Burch W Jr, Dalton S, Foresman TW, Rowntree R (1997) A conceptual framework for the study of human ecosystems in urban areas. Urban Ecosyst 1:185–199

Pickett STA, Cadenasso ML, Grove JM, Nilon CH, Pouyat RV, Zipperer WC, Costanza R (2001) Urban ecological systems: linking terrestrial ecological, physical, and socioeconomic components of metropolitan areas. Annu Rev Ecol Syst 32:127–157

Pickett STA, Belt KT, Galvin MF, Groffman PM, Grove JM, Outen DC, Pouyat RV, Stack WP, Cadenasso ML (2007) Watersheds in Baltimore, Maryland: understanding and application of integrated ecological and social processes. J Contemp Watershed Res Appl 136: 44–55

Pickett STA, Cadenasso ML, Grove JM, Groffman PM, Band LV, Boone CG, Brush GS, Burch WR Jr, Grimmond S, Hom J, Jenkins J, Law N, Nilon CH, Pouyat RV, Szlavecz K, Warren PS, Wilson MA (2008) Beyond urban legends: an emerging framework of urban ecology as illustrated by the Baltimore Ecosystem Study. BioScience 58:139–150

Pimentel D, Lach L, Zuniga R, Morrison D (2000) Environmental and economic costs of nonindigenous species in the United States. BioScience 50:53–65

Pouyat RV, Yesilonis ID, Nowak DJ (2006) Carbon storage by urban soils in the USA. J Environ Qual 35:1566–1575

Pouyat RV, Szlavecz K, Yesilonis I, Groffman P, Schwartz K (2010) Chemical, physical, and biological characteristics of urban soils. In: Aitkenhead-Peterson J (ed) Urban ecosystem ecology (Agronomy Monograph). ASA-CSSA-SSSA, Madison, WI (in press)

Pyšek P (1993) Factors affecting the diversity of flora and vegetation in central European settlements. Vegetation 106:89–100

Radeloff VC, Hammer RB, Stewart SI, Fried JS, Holcomb SS, McKeefry JF (2005) The wildland-urban interface in the United States. Ecol Appl 15:799–805

Rapoport EH (1993) The process of plant colonization in small settlements and large cities. In: McDonnell MJ, Pickett STA (eds) Humans as components of ecosystems: the ecology of subtle human effects and populated areas. Springer, New York, pp 190–207

Redman C, Grove JM, Kuby L (2004) Integrating social science into the long-term ecological research (LTER) network: social dimensions of ecological change and ecological dimensions of social change. Ecosystems 7:161–171

Reichard SH, Chalker-Scott L, Buchanan S (2001) Interactions among non-native plants and birds. In: Marzluff JM, Bowman R, Donnelly R (eds) Avian ecology in an urbanizing world. Kluwer, Boston, pp 179–223

Roy DB, Hill MO, Rothery P (1999) Effects of urban land cover on the local species pool in Britain. Ecography 22:507–515

Samu F, Szinetár C (2000) Rare species indicate ecological integrity:an example of an urban nature reserve island. In: Crabbé P, Holland AJ, Ryszkowski L, Westra L (eds) Implementing ecological integrity. Kluwer Academic, Dordrecht, pp 177–184

Sasvari L (1984) Bird abundance and species diversity in the parks and squares of Budapest. Folia Zool 33:249–262

Savard JL, Clergeau P, Mennenches G (2000) Biodiversity concepts and urban ecosystems. Landsc Urban Plan 48:131–142

Schaefer M, Kock K (1979) Ökologie der Arthropodenfauna einer Stadtlandschaft und ihrer Umgebung Laufkäfer (Carabidae) und Spinnen (Araneida). Schadlingskde Pflanzenschutz Umweltschutz 52:85–90

Schmid JA (1975) Urban vegetation: A review and Chicago case study, vol 161. Research Papers of the Department of Geography, University of Chicago, Chicago

Schwartz MW, Jurjavcic N, O'Brien J (2001) You can help rare plants survive in the cities. Nature 411:991–992

Schwartz MW, Jurjavcic NL, O'Brien JM (2002) Conservation's disenfranchised urban poor. BioScience 52:601–606

Schwartz MW, Thorne JH, Viers JH (2006) Biotic homogenization of the California flora in urban and urbanizing regions. Biol Conserv 127:282–291

Shapiro AM (2002) The Californian butterfly fauna is dependent on alien plants. Divers Distrib 8:31–40

Shochat E (2004) Credit or debit? Resource input changes population dynamics of city-slicker birds. Oikos 106:622–626

Shochat E, Warren PS, Faeth SH, McIntyre NE, Hope D (2006) From patterns to emerging processes in mechanistic urban ecology. Trends Ecol Evol 21:186

Slabbekoorn H, Peet M (2003) Birds sing at a higher pitch in urban noise. Nature 424:267

Smith RM, Gaston KJ, Warren PH, Thompson K (2006a) Urban domestic gardens (VIII): environmental correlates of invertebrate abundance. Biodivers Conserv 15:2515–2545

Smith RM, Thompson K, Hodgson JG, Warren PH, Gaston KJ (2006b) Urban domestic gardens (XI): composition and richness of the vascular plant flora and implications for native biodiversity. Biol Conserv 129:312–322

Sogge MK, Tibbitts TJ, Petterson JR (1997) Status and breeding ecology of the southwestern willow flycatcher in the Grand Canyon. Western Birds 28:142–157

Sorace A (2002) High density of bird and pest species in urban habitats and the role of predator abundance. Ornis Fennica 79:60–71

Steinberg DA, Pouyat RV, Parmelee RW, Groffman PM (1997) Earthworm abundance and nitrogen mineralization rates along an urban-rural land use gradient. Soil Biol Biochem 29:427–430

Sukopp H, Starfinger U (1999) Disturbance in urban ecosystems. In: Walker LR (ed) Ecosystems of disturbed ground (Ecosystems in the world). Elsevier Science, Amsterdam, pp 397–412

Sukopp H, Hejny S, Kowarik I (eds) (1990) Urban ecology: plants and plant communities in urban environments. SPB Academic, The Hague, pp 1–282

Swan CM, Pickett STA, Szlavecz K, Warren P, Willey T (2010) Biodiversity and community composition in urban ecosystems: coupled human, spatial and metacommunity processes. In: Holyoak M, Leibold MA, Holt RD (eds) Metacommunities: spatial dynamics and ecological communities. University of Chicago Press, Chicago (in press)

Szlavecz K, Placella SA, Pouyat RV, Groffman PM, Csuzdi CS, Yesilonis I (2006) Invasive earthworm species and nitrogen cycling in remnant forest patches. Appl Soil Ecol 32:54–63

Thompson HC (1999) Study finds adjacent land uses are key to predicting the number and type of exotic species in forest gaps (Maryland). Ecol Res 17:159–160

Tilghman NG (1987) Characteristics of urban woodlands affecting breeding bird diversity and abundance. Landsc Urban Plan 14:481–495

Troy AR, Grove JM, O-Neil-Dunne JPM, Pickett STA, Cadenasso ML (2007) Predicting opportunities for greening and patterns of vegetation and on private urban lands. Environ Manage 40:394–412. doi:10.1007/s00267-006-0112-2

Ulrich RS (1993) Biophilia, biophobia, and natural landscapes. In: Kellert S, Wilson EO (eds) The biophilia hypothesis. Shearwater/Island, Washington DC, pp 74–137

United Nations Population Fund (2007) State of the world population 2007: unleashing the potential for urban growth. United Nations Population Fund¸New York. http://www.unfpa.org/swp/2007/presskit/pdf/sowp2007_eng.pdf

US Census Bureau (2000) Census 2000; 1990 Census, Population and Housing Unit Counts, United States (1990 CPH-2-1)

USDA NRCS (U.S. Department of Agriculture, Natural Resources Conservation Service) (2001) 2001 Annual National Resources Inventory http://www.nrcs.usda.gov/technical/land/nri01/nri01lu.html

Velguth PH, White DB (1998) Documentation of genetic differences in a volunteer grass, *Poa annua* (annual meadow grass), under different conditions of golf course turf, and implications

for urban landscape plant selection and management. In: J Breuste J, Feldmann H, Uhlmann O (eds) Urban ecology. Springer, New York, pp 613–617
Vemuri AW, Grove JM, Burch WMA, Jr WR (2009) A tale of two scales: evaluating the relationship among life satisfaction, social capital, income, and the natural environment at individual and neighborhood levels in metropolitan Baltimore. Environ Behav. doi:10.1177/0013916509338551
Vilisics F, Hornung E (2009) Urban areas as hot-spots for introduced and shelters for native isopod species. Urban Ecosyst 12:333–345
Walker JS, Briggs BRC, JM KM, Warren PS, Wentz EA (2008) Birds of a feather: interpolating distribution patterns of urban birds. Comput Environ Urban Syst 32:19–28
Walton MB, Tsatiris D, Rivera-Sostre M (2006) Salamanders in forest-floor food webs: invertebrate species composition influences top-down effects. Pedobiologia 50:313–321
Wandeler P, Funk SM, Largiader CR, Gloor S, Breitenmoser U (2003) The city-fox phenomenon: genetic consequences of a recent colonization of urban habitat. Mol Ecol 12:647–656
Warren PS, Katti M, Ermann M, Brazel A (2006) Urban bioacoustics: it's not just noise. Anim Behav 71:491–502
Watson JEM, Whittaker RJ, Freudenberger D (2005) Bird community responses to habitat fragmentation: how consistent are they across landscapes? J Biogeogr 32:1353–1370
Watts CH, Lariviere MC (2004) The importance of urban reserves for conserving beetle communities: a case study from New Zealand. J Insect Conserv 8:47–58
Weiss MJ (2000) The clustered world: how we live, what we buy, and what it all means about who we are. Little, Brown, Boston
Whitmore C, Crouch TE, Slotow RH (2002) Conservation of biodiversity in urban environments: invertebrates on structurally enhanced road islands. Afr Entomol 10:113–126
Whitney G, Adams S (1980) Man as a maker of new plant communities. J Appl Ecol 17:431–448
Wilson EO (1984) Biophilia. Harvard University Press, Cambridge
Wilson EO (1987) The little things that run the world. the importance and conservation of invertebrates. Conserv Biol 1:344–346
Wolf JM, Gibbs JP (2004) Silphids in urban forests: diversity and function. Urban Ecosyst 7:371–384
Yamaguchi T (2004) Influence of urbanization on ant distribution in parks of Tokyo and Chiba City, Japan – I. Analysis of ant species richness. Ecol Res 19:209–216
Yeh PJ (2004) Rapid evolution of a sexually selected trait following population establishment in a novel habitat. Evolution 58:166–174
Yeh PJ, Price TD (2004) Adaptive phenotypic plasticity and the successful colonization of a novel environment. Am Nat 164:531–542
Ziska LH, Bunce JA, Goins EW (2004) Initial changes in plant population and productivity during secondary succession along an in situ gradient of carbon dioxide and temperature. Oecologia 139:454–458

Chapter 7
Indicators for Assessing Threats to Freshwater Biodiversity from Humans and Human-Shaped Landscapes

Robin Abell, Michele Thieme, and Bernhard Lehner

7.1 Introduction

There is little doubt that the activities of human populations, including their consumption of the products of freshwater ecosystems, have had appreciable impacts on freshwater ecosystems, the distribution of aquatic species, and the viability of those species' populations. Although data describing freshwater taxa are incomplete, ecologists know enough to conclude that, on average, these species are more imperiled than their terrestrial and marine counterparts (Allan and Flecker 1993; Williams et al. 1993; McAllister et al. 1997; Stein et al. 2000). Summary statistics from the IUCN (World Conservation Union) Red List show that, as of 2008, 27% of the world's freshwater amphibians and 16% of the world's freshwater crabs are classified as threatened (Collen et al. 2008; IUCN 2008). And although global data have yet to be fully collected and compiled for the other freshwater taxa, regional assessments (where available) find similarly alarming numbers of threatened species for these, as well. For example, 56% and 28% of freshwater fishes in the Mediterranean and East Africa, respectively, and 74% of Asian freshwater reptiles (Darwall et al. 2005; Smith and Darwall 2006; Turtle Conservation Fund 2002) are assessed as threatened by regional and compiled local studies. The projected mean future extinction rate of North American freshwater fauna – which lies within the range of estimates predicted for tropical rainforest communities (Ricciardi and Rasmussen 1999) – is approximately five times greater than the rate of terrestrial faunal extinctions and as much as three times the rate of coastal marine mammalian extinctions. Moreover, inventories of imperiled and extinct species can account only for described forms. Even within well-known groups such as fish, species are apparently becoming extinct before they can be properly classified (McAllister et al. 1985).

What underlies this loss? Decades of study suggest that a given species is rarely imperiled by a single stress. It is often very difficult to unravel the interrelated factors producing the disturbances and feedbacks that drive the depletion of native species within a watershed (Malmqvist and Rundle 2002). Miller et al. (1989) concluded that, of 40 extinctions among North American fishes, just seven could be narrowed to a single causative factor. In a more recent global analysis,

R.P. Cincotta and L.J. Gorenflo (eds.), *Human Population: Its Influences on Biological Diversity*, Ecological Studies 214,
DOI 10.1007/978-3-642-16707-2_7,

Harrison and Stiassny (1999) determined that 71% of fish extinctions were related to habitat alteration, 54% to the introduction of exotic species, 26% to pollution, and the rest to hybridization, parasites, diseases, or intentional eradication. The array of stresses besieging freshwater species and their habitats include dams, exotic species, over-fishing, pollution, stream channelization, climate change, water withdrawals and diversions, and the panoply of impacts from development of the terrestrial landscape. Many, but not all, of these threats can be traced – at least in part and very often indirectly – to pressures associated with nearby settlement, with increasing local population density, or distant human population growth.

This essay focuses on broad-scale threat assessments, with specific reference to human population density and distribution as assessment tools. This topic was chosen because there is growing consensus that planning at broad scales best enables the realization of fundamental biodiversity conservation goals [e.g., the maintenance of viable species populations within resilient blocks of natural habitat (Noss 1992; Groves et al. 2002)]. Many abiotic and biotic processes (e.g., seasonal flooding, riparian plant-species succession, reproductive migration) essential for the maintenance of freshwater biodiversity operate at broad scales. As background to our discussion, we provide a brief overview of threats to freshwater systems referring readers to the rich literature on the topic for more in-depth treatment (also see Dudgeon et al. 2006) and review the growing literature on the use of landscape indicators to assess aquatic ecological integrity (see Gergel et al. 2002a).

Following these reviews, we focus on threat assessments over large areas in which limited data are available, as is the case in much of Asia, Africa, and Central and South America. Whereas most aquatic biodiversity assessments have been undertaken in the data-rich temperate world, particularly in study areas covering mid- to small-size catchments, we illustrate approaches taken at the global, continental (Africa), and large river-basin (Niger) levels. And we conclude – revisiting the link between human populations and impacts on freshwater species – by discussing research priorities and the need to address key gaps in data that hinder assessment and ultimately impede managers' abilities to mitigate threats.

7.2 Threats to Freshwater Biodiversity

For the purposes of assessing threats to freshwater species and habitats, we place the universe of threats in three broad categories: (a) activities in the watershed's terrestrial realm (e.g., deforestation, agriculture, land use change, and road building) resulting in altered flow regimes, water quality, and physical habitat attributes, (b) point source pollution, disturbances that occur directly in aquatic, wetland, or riparian environments (e.g., dams and channelization), and (c) threats that affect freshwater organisms directly (e.g., exotics and overexploitation) (Abell et al. 2000). A general list of common threats in each category can be used to catalyze discussions about threats that are specific to particular freshwater systems (Table 7.1). Response to disturbance, however, can vary considerably with freshwater geomorphology.

Table 7.1 Examples of general threats to aquatic species and habitats [modified from Abell et al. (2002)]

Threats
Terrestrial threats (land cover change)
Logging and associated road building
Grazing, particularly in riparian zone
Agricultural expansion and clearing for development
Urbanization and associated changes in runoff
Mining or other resource extraction and associated road building
Aquatic threats (direct habitat modification)
Degraded water quality (e.g., point or nonpoint source pollution; changes in temperature, pH, dissolved oxygen (DO), other physical parameters; sedimentation and/or siltation; salinity)
Altered hydrographic integrity (flow regimes, water levels) resulting from dams, surface or groundwater withdrawals, channelization, etc.
Habitat fragmentation from dams or other barriers to dispersal
Reduced organic matter input
Additional habitat losses, such as siltation of spawning grounds
Excessive recreational impacts
Biotic threats
Unsustainable fishing or hunting
Unsustainable extraction of wildlife or plants as commercial products
Competition, predation, infection, and genetic contamination by exotic species
Genetic effects of selective harvesting
Intentional eradication

And, although most threats originate within a system's watershed, some – such as acid deposition and climate change – originate beyond watershed boundaries. These extra-watershed stresses can be strong enough to alter the timing, duration, magnitude, and direction of hydrologic processes, including groundwater flows, shifts in water temperature, and changes in riparian vegetation and related organic matter inputs (Poff et al. 2002). Although these categories are simply organizational tools that can be reconfigured in numerous ways, most categorizations of freshwater threats by other authors are similar (see O'Keeffe et al. 1987; Allan and Flecker 1993; Boon 2000; GIWA 2002; Malmqvist and Rundle 2002).

Not all threats are created equal. Within the terrestrial realm, certain anthropogenic activities stand out. The conversion of natural land cover to impervious surfaces (paved roads, parking lots, buildings) can have substantial effects on hydrology (Jones et al. 2000). Unpaved roads, often constructed in association with logging, at a minimum can increase sedimentation and be the source of slope failures (Sedell et al. 2000; Sidle et al. 2004). Industrial and urban activities have long been associated with point source pollution, and agriculture is implicated in various forms of nonpoint-source (or diffuse) pollution, leading to problems including eutrophication, poisoning, siltation, and sedimentation (Allan and Flecker 1993). Agriculture is also the largest user of abstracted fresh water (Revenga et al. 2000). Within each of these categories of threat, there is a high degree of variation. For example, paving over the most hydrologically active lands within a watershed – those contributing the most to downstream flow – will have a greater effect on hydrology than paving over other lands. The construction of logging roads on high-gradient slopes is more likely to lead to significant erosion than will road construction on flatter terrain. No-till, pesticide-free agriculture typically generates less nonpoint-source pollution than other methods. And waste from industrial and urban sources can create major or minor impairments, depending on the constituent materials and the volume of effluent.

Direct disturbances to the aquatic environment are considered among the most destructive. In nearly all types of freshwater ecosystems, dams create barriers to the movement of organisms and alter downstream and upstream habitats in a variety of ways (Ligon et al. 1995; World Commission on Dams 2000). However, the type and operational scheme of a dam or run-of-the-river station, its location, and the qualities of the river itself can produce substantial mediating effects (Poff and Hart 2002). In floodplain river systems, levees, dams, and other anthropogenic structures built to prevent flooding can deprive species of important feeding, spawning, and nursery habitats (Gergel et al. 2002b; Tockner and Stanford 2002). The loss of riparian forests in headwater streams can have major repercussions for organic matter inputs and water temperature regulation (Gregory et al. 1991; Peterson et al. 2001; Saunders et al. 2002). Water withdrawal is another serious direct disturbance to freshwater ecosystems around the world, particularly in arid and semiarid environments (Postel 2002). Withdrawals from rivers and lakes often occur with little or no regard for ecosystem water requirements.

Among direct threats to native freshwater biota, the introduction and propagation of nonnative species has been identified as among the most serious. However, other activities can impact native communities at similar magnitudes. Small-mesh nets and other unselective fishing methods can deplete populations of a range of taxa, simultaneously selecting against the survival of larger individuals and species (Allan et al. 2005; Welcomme 2005). Targeting fish while they are breeding or in refuge areas can reduce the viability of populations (Welcomme 2001; Cowx 2002). But not all forms of exploitation are necessarily unsustainable and some conservationists have promoted sustainable harvests – including the harvest of fish species used as ornamentals – as an alternative to more harmful forms of exploitation (Chao and Prang 1997).

Population density is typically an effective predictor of several types of disturbances to aquatic ecosystems. At the river-basin level, population density is statistically associated with nitrogen and phosphorus pollution from urban and agricultural runoff (Turner et al. 2003) in that basin, although these relationships are not equally strong across all river sizes (Caraco et al. 2003). In large basins, population size is highly correlated to water withdrawals, a relationship largely attributable to demand for food and irrigation (Gleick 2000). However, the vast majority of global population growth is occurring (and will continue to occur) in urban areas, where industrial and urban pollution, altered runoff, and loss of riparian vegetation may pose greater threats to the rivers on which many cities are built than do water withdrawals (Malmqvist and Rundle 2002).

In general, patterns of human-altered landscapes offer more comprehensive information about specific threats to freshwater communities than does a simple assessment of nearby human population density. However, although researchers have amply demonstrated that the loss of watershed-wide terrestrial vegetative cover negatively affects the integrity of freshwater ecosystems, ecologists are only beginning to understand the more nuanced impacts of the location and configuration of converted land on aquatic biota (Gergel et al. 2002a). Regional studies have identified relationships between biotic integrity and impervious cover, cropland, and other landscape indicators. But the general conclusions of individual studies are often at odds with one another. Frequently, the lack of consensus can be attributed to differences in the scale at which research was conducted.

Despite the accomplishments of recent research, some of the most fundamental questions concerning anthropogenic impacts on aquatic biodiversity have yet to be fully answered. Given the basic theoretical and methodological shortcomings, plus the gaps in data, how should aquatic ecologists proceed when they are charged with the responsibility of identifying and assessing threats to freshwater biodiversity?

7.3 Direct and Indirect Evaluations of Threats to Aquatic Biodiversity

Methods that directly evaluate the degree of threat to freshwater biodiversity in an aquatic ecosystem typically utilize species composition and abundance data. These data require extensive sampling, expertise, and effort, and thus the status of only a very small fraction of the world's freshwater species has been evaluated using direct methods. Amphibians represent the best-studied aquatic group, other than water birds (IUCN 2001; Stuart et al. 2004). Of the approximately 44,000 scientifically described freshwater species, about 11,000 – the distributions of which are only partially mapped (Reaka-Kudla 1997) – are considered in the 2008 IUCN Red List and about 2,200 of those are judged as data deficient (IUCN 2008). Without adequate knowledge of species distributions and their degree of imperilment, this approach cannot uniformly identify those freshwater systems under highest threat. Direct evaluations of loss can also produce misleading results when time lags occur between the introduction of a stress and its effects on habitat quality and native-species population dynamics (Harding et al. 1998).

Approaches that indirectly assess the biotic community evaluate, instead, the ecological integrity of aquatic ecosystems. Karr and Dudley (1981, p. 56) define ecological integrity as an area's "capability of supporting and maintaining a balanced integrated adaptive community of organisms, having a species composition, diversity and functional organization comparable to that of natural habitat of the region" – in effect the inverse of the level of threat (for a debate over definitions and measurement, see De Leo and Levin 1997; Boulton 1999; Norris and Thoms 1999). Assessments of ecological integrity make use of one or more multimetric indices that incorporate data on water chemistry, instream flow, physical habitat, biotic features, or a combination thereof (Gergel et al. 2002a). These techniques are typically resource-intensive and time-consuming and often problematic in river systems in developing countries that are extensive and inaccessible. In some cases, acceptable reference sites may simply not exist.

For broad-scale, time-constrained, resource-limited conservation planning, the most appropriate tool for assessing the impairment of freshwater systems is often a fairly straightforward, relativistic assessment of ecological integrity using various types of geospatial data (O'Neill et al. 1997) (Table 7.2). Data characterizing the extent of agriculture in a watershed or the size and location of dams – measures

Table 7.2 A list of geospatial data layers that could serve to inform an evaluation of ecological integrity [modified from Abell et al. (2002)]

Biotic
Vegetation/land cover
Indigenous areas
Cattle/livestock densities
Aquaculture operations
Human population density
Areas of deforestation
Ranges of exotic species, or areas of known introductions
Abiotic
Roads
Towns and cities
Land uses (current and historic)
River network (e.g., derived from Digital Elevation Model)
Runoff (by grid cell)
Discharge (by river segment)
Erosion potential (by grid cell)
Sediment transfer (by grid cell)
Water quality
Water temperature
Extent of floodplains
Fishing centers
Industrial sites
Major ports
Railroads
Pipelines (present and planned)
Mining activity and concessions
Toxic sites
Logging activity and concessions
Irrigated and nonirrigated croplands
Pesticide application
Refineries
Power generation plants
Interbasin water transfers (present and planned)
Water abstractions/Water use
Channelized or diked streams
Canals
Drainage projects
Impoundments and reservoirs (present and planned), plus additional barriers to passage
Fish passage devices (working and failing)
Areas of conflict
Protected areas

referred to as "proxies" – can provide insights into the relative degree of system stress.

Some proxies are less removed from actual measures of ecological integrity than others. For example, mapping the large dams in a river system allows for a rough calculation of the amount of riverine habitat that has been disturbed both above and below the dams, as well as habitat that has become functionally lost to species unable to traverse the dams. The percentage of converted land within a watershed can also serve as a coarse proxy for the complex of hydrologic and biogeochemical changes resulting from conversion (Hughes and Hunsaker 2002). Such an analysis, by itself, does not generally provide sufficient information to identify a solid land-conversion threshold above which managers should expect significant threats to freshwater species, although there is evidence that such thresholds exist (Wang et al. 1997). However, the method provides a basis for comparisons among similar watersheds.

7.4 Large-Scale Threat Analyses

Large-scale assessments of threats are necessarily coarse; their data inputs are rarely, if ever, of uniform quality, and they may feature bias and sampling gaps. Yet, despite their limitations, large-scale assessments such as global evaluations can illustrate spatial and, less often, temporal patterns, providing powerful tools for setting priorities, raising awareness, and catalyzing action. For large-scale studies, the level of analysis is often prescribed by the resolution of available data. Published sources often provide only single estimates for large river basins or a country-level value (e.g., human water-use data). Spatial interpolation algorithms can be employed to improve resolution, but add uncertainty. Point data – such as the coordinates of dams or gauging stations (the latter can be obtained from the Global Runoff Data Center in Koblenz, Germany) – can be used for spatially explicit analyses, but these data vary globally in quality and completeness.

7.4.1 Global-Scale Analyses

The most comprehensive global vector maps of hydrographic features (rivers, lakes, wetlands) are typically in the range of 1:1,000,000–1:3,000,000 resolution (e.g., Environmental Systems Research Institute 1992; Environmental Systems Research Institute 1993; Lehner and Döll 2004) and are available as country, regional, or global coverages (e.g., Hearn et al. 2000–2003). Global assessments characteristically use grid cells as their unit of analysis, with typical cell sizes ranging from 0.5° to 30 seconds (about 50–1 km). Some global data, such as human population densities, are provided in several resolutions and data formats (both vector and grid) or are the product of postprocessing algorithms. The

Gridded Population of the World (Center for International Earth Science Information Network and Centro Internacional de Agricultura Tropical 2005) uses, as its source of quantification, census data that are associated with administrative units. In contrast, the LandScan database (Oak Ridge National Laboratory 2009) provides geospatial population counts that have been adjusted by redistributive modeling, shifting human distribution toward certain land cover types and areas adjacent to roads. Each has its advantages and disadvantages, depending on the application.

For freshwater threat assessments, the most meaningful level of analysis is generally the drainage basin and its subbasins. The extent of drainage basins can be determined from digital elevation models, which exist as global coverages in various resolutions. Geographic Information System (GIS) technology offers the possibility to identify the upstream catchment for any individual location within a river system, allowing for the division of subbasins at different scales.

Since the Shuttle Radar Topography Mission (SRTM) elevation data were released at 90-m resolution at near global coverage (Farr and Kobrick 2000), significant advancements have been made toward a new generation of drainage maps. The HydroSHEDS database (Hydrological data and maps based on Shuttle Elevation Derivatives at multiple Scales), developed by World Wildlife Fund (WWF) (Lehner et al. 2008), provides a suite of data layers and hydrographic information at various resolutions (including 90 m, 500 m, and 1 km) and allows scientists and managers to use preproduced river maps and watershed boundaries or to create their own digital products. Additional attribute layers are in progress, including discharge estimates and river orders. In practice, however, even the best elevation data sets and their derived products are afflicted with errors and uncertainties, particularly in flat topographies, such as large floodplains or in coastal zones. Moreover, the radar-based SRTM elevation data and its derivatives (including HydroSHEDS) are influenced by terrestrial vegetation, and the SRTM coverage does not extend above 60°N latitude.

The most appropriate basin size in any particular analysis is not necessarily obvious. Because data are often unavailable or too coarse for smaller basins, most global threat assessments default to using only the world's largest river basins. However, large river basins are not necessarily the most appropriate units for freshwater biodiversity conservation planning. Such planning typically focuses on protecting representative and distinct species assemblages, which necessitates incorporating biogeographic information into the delineation of planning units (Groves 2003). A large river basin may encompass sufficiently sharp biogeographic discontinuities to warrant dividing it for the purposes of planning. Or it may include multiple basins that are so similar that they can be melded into a single planning unit.

In part to address the need for biogeographically informed planning units, WWF and The Nature Conservancy (TNC) have developed a global map of "freshwater ecoregions." A freshwater ecoregion is defined as a large area, encompassing one or more freshwater systems, that contains a distinct assemblage of natural freshwater communities and species. The freshwater species dynamics and environmental conditions within a given ecoregion are more similar to each other than to those species dynamics and environmental conditions that characterize surrounding ecoregions. Generally, freshwater ecoregions comprise adjacent watersheds that are aggregated on the basis of freshwater zoogeography, with an emphasis on fish species (Abell et al. 2008). The WWF/TNC ecoregions cover all freshwater systems and provide biologically relevant units for future broad-scale threat assessments. Ultimately, the target management authority, or decision-maker, may determine whether recommendations are framed in terms of basins or ecoregions.

Existing global analyses of freshwater biodiversity and their threats are rare. One of the most comprehensive assessments of condition of the world's freshwater systems is the World Resources Institute's (WRI) *Pilot Analysis of Global Ecosystems* (PAGE) (Revenga et al. 2000; also see the Global Water System Project 2005). Although PAGE's indicators of condition are focused on the maintenance of ecosystem services, rather than on biodiversity, many of these measures are broadly applicable as proxies for aquatic ecological integrity. The PAGE analysis compiles the results of multiple studies and presents continental and global maps evaluating threats in more than 200 large river basins. Below, we summarize several of the main PAGE analyses and present examples of other recent global studies that examine similar subjects, grouping the analyses using the three categories of threats (presented in Table 7.1).

Global analyses of aquatic habitat threats using geospatial data have typically focused on dams. Threats are associated with aspects of main-channel and tributary impoundment, reservoir storage volumes in relation to natural river flow, and alterations of seasonal patterns of discharge (Revenga et al. 2000). On the assumption that water and sediment retention result in a wide range of impacts to normally free-flowing river ecosystems, Vörösmarty et al. (1997) estimated the change in residence time from data on more than 600 large reservoirs in 236 river basins around the world. The results demonstrated a dramatic aging in river runoff, suggesting strong biophysical alterations to these systems. Dynesius and Nilsson (1994) and Nilsson et al. (2000, 2005) investigated additional flow criteria to develop a river fragmentation index for large river systems. By synthesizing data on dams, reservoirs, interbasin transfers, and irrigation consumption (irrigation water not returned to a river), they determined that 77% of the total discharge of the 139 largest river systems in the northern third of the world is strongly to moderately affected by the combination of these factors. This fragmentation could profoundly affect species populations over a substantial area of the world (Rosenberg et al. 2000).

Another indicator that has been applied to evaluate the magnitude of anthropogenic aquatic habitat threats is the level of water stress, which has been measured as the water withdrawals-to-availability ratio (WTAR). As an example, we present results of the WaterGAP2 model (Döll et al. 2003; Alcamo et al. 2003), which provides discharge and human water use calculations for the present time and for future scenarios on a global 0.5° grid (Fig. 7.1). The higher the WTAR, the more intensively river-basin water is used, and therefore (we assume) the more stress is placed on the freshwater ecosystem. Thresholds have been applied to identify areas of *low water stress* ($0.1 < \mathrm{WTAR} \leq 0.2$), *medium water stress* ($0.2 < \mathrm{WTAR} \leq 0.4$), and *high water stress* ($\mathrm{WTAR} > 0.4$) (UN 1997).

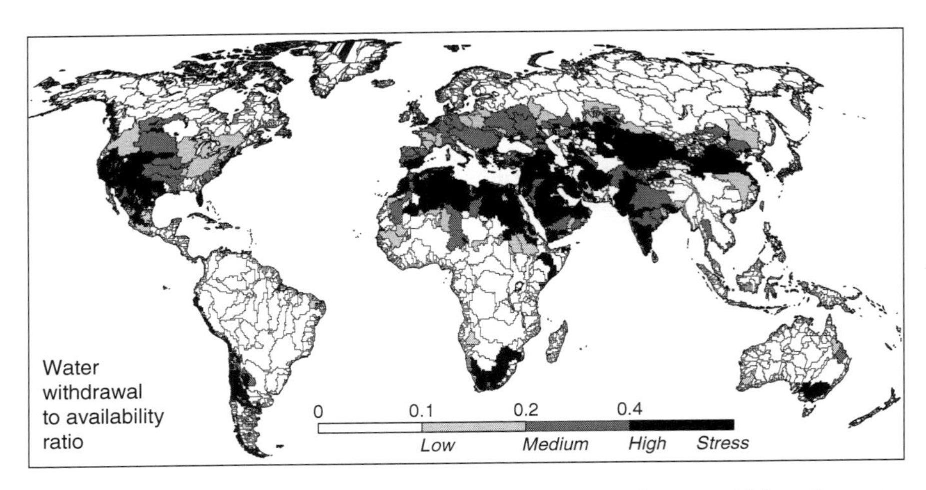

Fig. 7.1 Water stress, expressed as the ratio of annual anthropogenic water withdrawals to water availability (WTAR), among Earth's river basins. The values are based on results of the Water-GAP2 model (Döll et al. 2003; Alcamo et al. 2003). Values were aggregated from grid cells at 0.5° resolution to basins and subbasins to account for uncertainties in the location of water withdrawals and water availability

However, the significance of these thresholds has not been conclusively verified. Moreover, these indicators do not take ecosystem needs into account and few studies included environmental flow requirements into their assessments (e.g., Smakhtin et al. 2004). To capture interannual variability and seasonal fluctuations in water stress, additional measures for critical low flows like the Q90 (discharge that is exceeded at least 90% of time), or the duration and severity of broad-scale droughts, should be assessed.

In the category of direct threats to aquatic biota, the PAGE analysis offers two global measures: country-by-country fish catch trends, annually from 1984 to 1997 [United Nations Food and Agriculture Organization (FAO) 1999]; and documented introductions of fish and invertebrate species, compiled by the FAO (1998). Both are presented as an indicator of impaired ecosystem services. Because overexploitation is ultimately a threat that can be addressed through regulation, country-level fish catch data can be useful. However, neither the catch data nor the introduced-species data indicate which river or lake systems are experiencing reduced catches or increased introductions, and they do not suggest why these occurred.

In addition, two other PAGE indicators reflect disturbances to the terrestrial realm that could be transmitted to aquatic systems. The percentage of cropland in major river basins is estimated from the Global Land Cover Characterization (US Geological Survey et al. 1997) at 1-km resolution. Percentage of urban and industrial use in major river basins is derived from the US National Oceanographic and Atmospheric Administration-National Geophysical Data Center's (1998) Stable Lights and Radiance Calibrated Lights of the World, 1994–1995. Similar analyses could be conducted for other aggregate units, such as subbasins or freshwater ecoregions. More recent coverages, including the Global Land Cover 2000 database

(Joint Research Centre and European Commission 2003) and the Moderate Resolution Imaging Spectroradiometer (MODIS) land cover data set (Hodges et al. 2001), may allow improved accuracy. Analyses combining land cover data with other global maps are also possible (Table 7.3). Nonetheless, two common factors associated with remote-sensing-derived land cover data introduce significant uncertainty: variation in interpretation and the lack of an internationally accepted land cover classification. Given that different source data and classifications can generate different results, analyses of land cover data sets are coarse and should be interpreted with care.

The extensive collection and digitization of global environmental data has prompted efforts to update the Ramsar Wetlands Database (Wetlands International 2007), which provides information on putative protection to important freshwaters and wetlands. An enhanced global map of irrigated areas (Siebert et al. 2005) permits improvements to fragmentation analyses as well as calculations of the proportion of irrigated areas in basins and ecoregions. Making use of the most recently developed global data sets and methodological advancements, the study by Vörösmarty et al. (2010) arguably provides the most explicit and spatially detailed assessment of global threats to river biodiversity to date, as well as access to their underlying databases.

Table 7.3 A list of specific geospatial indicators that could be used in models assessing aquatic ecological integrity

Percentage of land use classes, by subbasin (e.g., 20% forest, 40% agriculture, 10% urban, etc.)
Average forest (or other native cover) patch size as a percentage of subbasin area, index of forest connectivity, or measure of dominance of land uses
Percentage of land use classes within fixed-width buffer of streams or other water bodies
Connectivity of riparian forest (or other native cover type)
Length or percentage of streams with riparian vegetation cover, by subbasin
Road density or number of road-stream crossings, by subbasin
Sediment contribution or erosion potential, by subbasin
Potential nutrient export, by subbasin
Average discharge, flow accumulation, or runoff of grid cells, by subbasin
Urban expansion or population growth, by subbasin
Percentage of area grazed, by subbasin
Percentage of headwaters (defined by elevation, gradient, stream order, etc.) with original land cover, by subbasin
Average population density, by subbasin
Degree of protected area coverage (all areas, or only aquatic habitats), by subbasin
Number or coverage of mining, logging, or other resource extraction operations, by subbasin
Number of pipeline-stream crossings, or length of pipeline, by subbasin
Number of impoundments per stream length, by subbasin
Length of stream flooded by impoundments, or length of stream above impoundments made inaccessible to migrating species, by subbasin
Number or length of free-flowing streams, divided by number or length of impounded streams
Length of stream habitat lost as a result of channelization (requires historic and current stream morphology maps)
Length or area of floodplain habitat cut off from river
Flow modification

For a number of indicators, however, worldwide geospatial coverage remains incomplete. For example, there is a comprehensive database of the world's large dams (>15 m high) and their reservoirs (International Commission on Large Dams 2007), but geographic coordinates are not provided. Efforts to produce a geo-referenced Global Reservoir and Dam database (GRanD) are currently underway (Lehner et al. 2010). There are still no detailed global maps of changes in groundwater resources, changes in the extent of wetlands, changes in water quality, or sediment fluxes (Revenga et al. 2000). Spatial data that capture these temporal dynamics are available only for specific rivers, regions, and countries. For example, FAO's Land and Water Development Division provides a database on sediment yields for many, but not all, rivers worldwide (FAO 2001); and detailed data on dams are provided in the National Inventory of Dams of the United States (US Army Corps of Engineers 2007).

7.4.2 Continental-Scale Analyses

Although data availability and quality constraints continue to pose limits, an increasing number of continental-scale research projects are generating new and more complete regional data sets. FAO's Africover Project is establishing a digital geo-referenced database identifying land cover, roads, and hydrography for all of Africa at a scale of 1:200,000 (see first ten country sets, FAO 2003). Other germane examples include: the Africa Data Sampler (World Resources Institute et al. 1995); data on human population change in Africa from 1960 to 2000 (United Nations Environment Programme/Global Resource Information Database et al. 2004); and FAO's database of African dams and reservoirs (FAO 2006).

Expert assessments, based on biologists' first-hand knowledge of aquatic habitats and biota, can enhance, edit, and validate the results of geospatial analyses. Expert assessment is typically introduced collectively in a workshop setting, or individually through consultations (Groves 2003), and conservation planners and managers increasingly turn to traditional ecological knowledge of local people to supplement and inform information derived from academics and field biologists (Berkes et al. 2000). At the continental scale, expert assessment can include a range of biases toward certain study sites and taxonomic groups, which might occur with particular sampling techniques.

In assessing threats to Africa's freshwater ecoregions and their biodiversity, Thieme et al. (2005) combine expert assessment and an evaluation of digital data. To assess land-based threats, the authors merge land cover and population data to determine the percentage of degraded land within each ecoregion. Areas classified as urban and cropland are considered degraded from a freshwater biodiversity perspective, as are areas with population densities $\geq$10 people km^{-2} (Fig. 7.2). Based on the percentage of degraded land, ecoregions are assigned a threat level of low, medium, or high (Fig. 7.3a). Authors typically employ local-expert assessments obtained using surveys, combined with literature reviews, to validate the

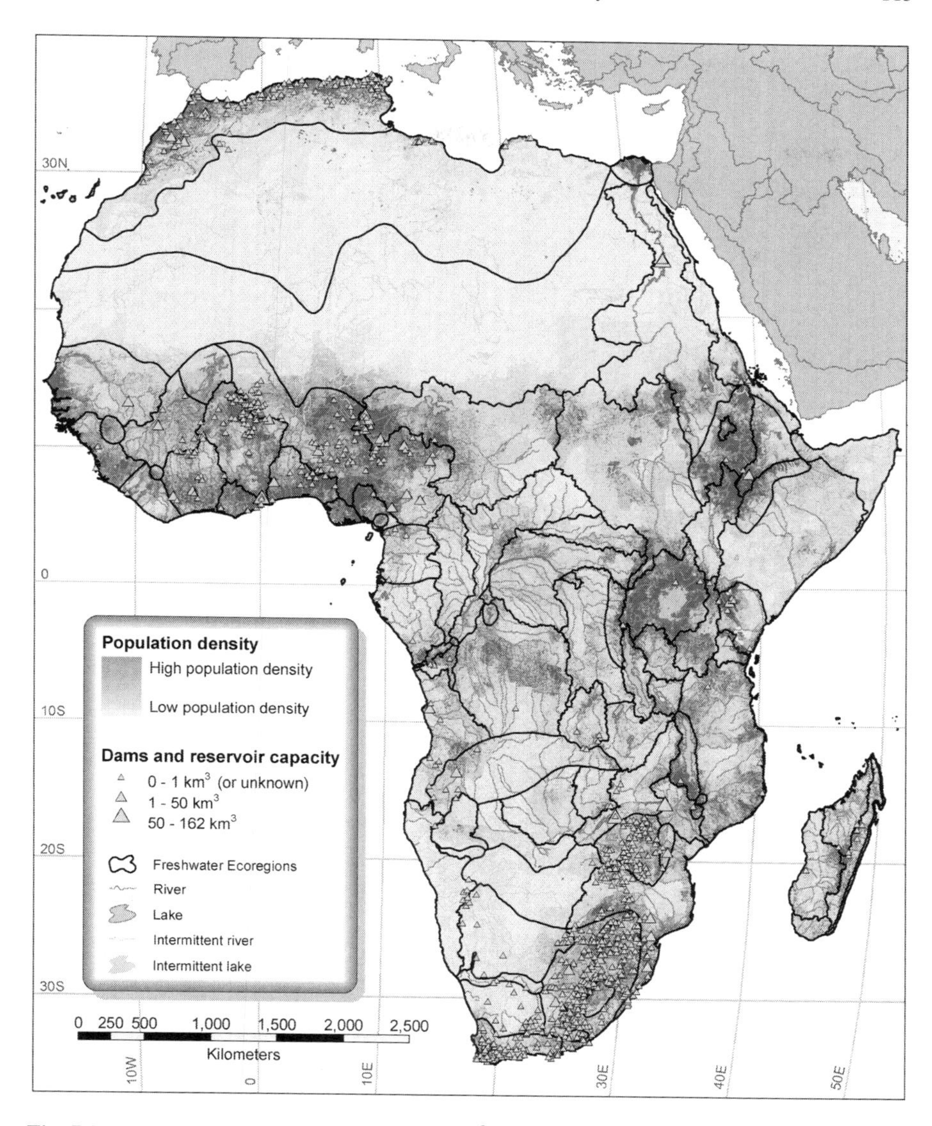

Fig. 7.2 Human population density (people km^{-2}) for the year 2000 and major dams within the freshwater ecoregions of Africa [from Thieme et al. (2005), population data from ORNL (2001), dam data from FAO (2002)]

results of map-based analyses. Where experts disagree with the map-generated threat levels, evidence from expert assessment is usually given greater weight.

In a second analysis, Thieme et al. (2005) evaluate direct threats to aquatic habitats, relying first on expert assessment and literature. They then determine which ecoregions contain dams greater than 15 m in height (see Fig. 7.2), elevating those ecoregions with more than 50 dams greater than 15 m high to the highest

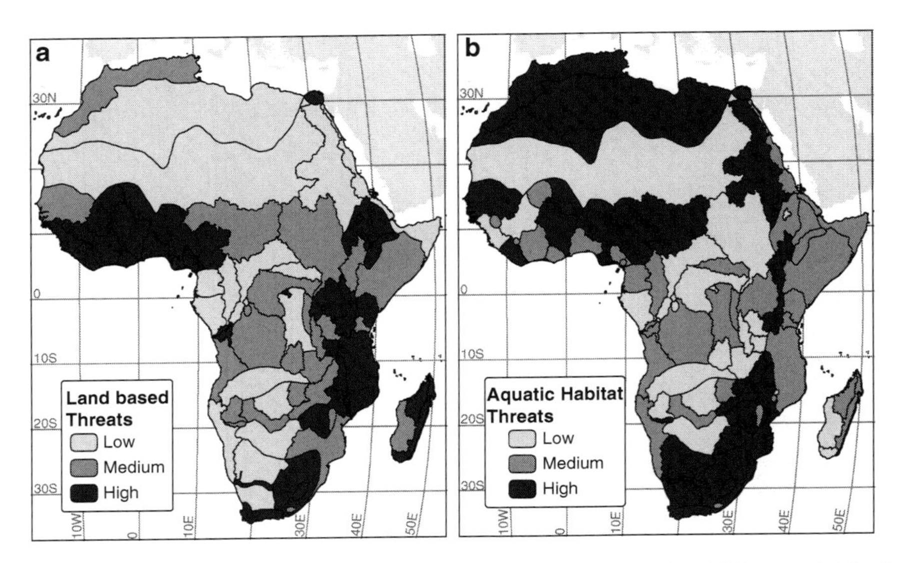

Fig. 7.3 Threats to the freshwater ecoregions of Africa and Madagascar, from Thieme et al. (2005) (**a**) Land-based threats as assessed through an analysis of degraded land cover (cropland and urban areas) (Hansen et al. 2000) and population ($\geq$10 people km^{-2}) (ORNL 2001), and expert opinion; (**b**) Aquatic habitat threats as assessed by expert opinion and through literature review, as well as through regional analyses of numbers of dams and reservoir capacity (FAO 2002)

threat level. Those ecoregions with fewer dams, but an extremely large reservoir capacity (>150 billion m^3), are also elevated to this threat level (see Fig. 7.3b).

Land-based threats were assessed to be high in all of West Africa, in parts of southern and East Africa, much of Madagascar and nearby islands, several Atlantic coastal islands, in the Rift Valley, and in the Ethiopian highlands. In the vast majority of ecoregions, expert assessment coincided with analyses of geospatial data. In only 7 of 93 cases, experts evaluated the level of threat to be higher than the level suggested by geospatial data. Aquatic habitat threats were assessed as high over large portions of the African continent including parts of North and West Africa, as well as a large block of ecoregions in southern Africa. Among those with a ranking of "high," seven ecoregions contained more than 50 dams. In general, large numbers of dams tend to occur in areas experiencing high human population densities.

7.4.3 River-Basin Threat Analyses

Narrowing the scope of analysis to the river-basin scale has many advantages. At this level, employing high-resolution remote sensing imagery is often feasible, both from a financial and workload perspective. The narrow scope tends to give analysts the opportunity for greater data manipulation and manual corrections.

Within a relatively small study area, gauging stations and dam locations may be reliably linked to river reaches. Rivers, lakes, and wetlands can be aligned and topologically connected. And species distributions and biodiversity priority areas can be delineated with enough accuracy to identify the main tributaries draining into them. Rather than settling for coarse proxy indicators to score threats, analysts can apply finer-scale relationships to discern the nature of threats to habitat and species. For example, rather than employing a count of the total number of dams per ecoregion as a threat indicator, at the river-basin level, it is possible to derive the distance between dams, to distinguish dams on the main river course from those on headwater tributaries, to identify hydrologically sensitive areas, and to separately assess threats originating upstream and downstream of areas of particular interest within the basin.

In the following section, we review several analyses of threats to the freshwater biodiversity of the Niger River Basin to illustrate the kinds of analyses and level of detail that are possible for a relatively data-poor large river system. In April 2002, the Niger Basin Initiative (NBI) – a partnership between WWF, Wetlands International, and the Nigerian Conservation Foundation – hosted a workshop at which some 40 biologists, ecologists, and hydrologists from seven countries within the basin mapped the areas most critical to freshwater biodiversity conservation (Wetlands International 2002; Niger Basin Authority 2007). Areas of biological significance were selected based on expert assessment, which was informed by available biodiversity data for fish, birds, and other vertebrates. Additionally, data quantifying runoff generation, produced by a global hydrological model (Döll et al. 2003), were prepared for the Niger River Basin and served as a starting point for discussing which subbasins were most important for maintaining the flow regime. By combining these data, experts selected 19 priority areas for long-term freshwater biodiversity conservation (Fig. 7.4). Threat analyses were then focused on these priority areas. Using GIS, human population densities were calculated (1) within the priority area and (2) in the associated upstream watershed for each priority area (Fig. 7.4).

At this scale of analysis, patterns of human settlement and their population density in the Niger Basin are relevant to conservation planning. Average population densities were high (21–50 people km^{-2}) in a majority of the 19 priority areas and were highest (>50 people km^{-2}) in the Niger Delta and a tributary to the Benue River – two areas located in the lower portion of the Niger Basin. Upstream analyses underscore the complexity of interpreting freshwater threats. For priority areas in the middle and lower Niger, upstream human population density levels tended to be low, reflecting sparsely populated arid expanses of Sahel and Saharan terrestrial ecosystems in their expansive, but seasonally dry catchments. However, as these upstream areas contribute insignificant discharge, the low population density values cannot be interpreted as an indicator of low human threats, and absolute population numbers cannot be discounted (e.g., total upstream population from the Niger Delta is 95 million people, as calculated from the LandScan database, Oak Ridge National Laboratory 2009). Among areas that had a higher human population density upon inclusion of upstream tributaries was the Yankari

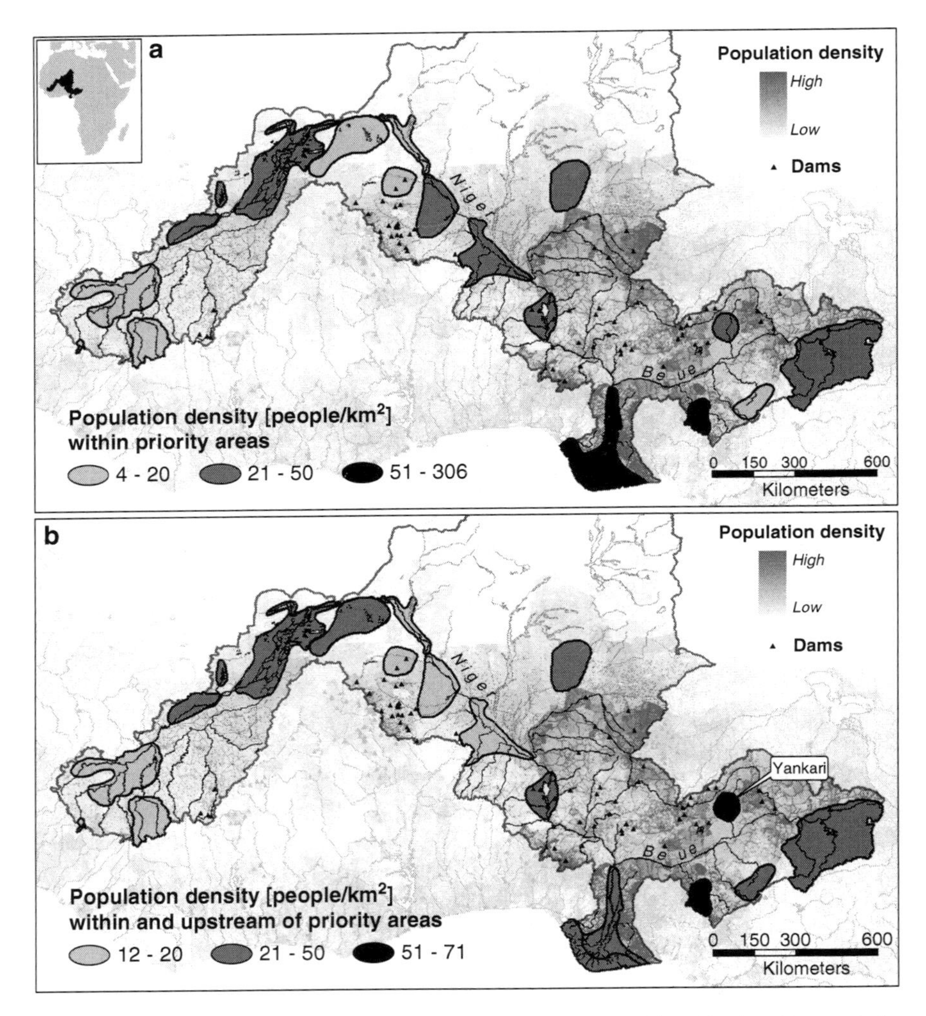

Fig. 7.4 Priority areas for freshwater conservation in the Niger River Basin with population density (people km^{-2}) within each area (**a**), and population density within each area and its upstream catchment (**b**). Freshwater priority areas identified at the Niger Basin Initiative's priority setting workshop in April 2002 (Wetlands International 2002). Population data are from ORNL (2001); dam data from FAO (2002)

National Park priority area in the Benue catchment with >50 people km^{-2} in its upstream catchment as compared to 21–50 people km^{-2} within the priority area alone. Clearly, in cases like the Yankari National Park, conservation strategies aimed at limiting biodiversity-affecting human activities will need to focus not only within the priority area, but also in the upstream catchment area – a basic principle of integrated river-basin management.

7.5 Conclusions

The ability of managers to mediate impacts on freshwater species and their habitats is, in large part, dependent on ecologists' ability to identify the type and magnitude of the threats that currently place them among the most imperiled elements of global biodiversity. What do decades of aquatic ecosystem research tell us about how the immediate presence of human population figures among today's threats to freshwater biodiversity? Areas with high human population numbers generally mean degradation of aquatic ecosystems. However, the opposite is not always true. The type and magnitude of stresses to aquatic systems are mediated by the type of settlement, and there is often a spatial disconnect between the location of dense concentrations of humans and their impacts (e.g., hydropower dams and electricity users, irrigation and food consumers, loggers and home builders, fishermen and seafood consumers). In addition, the moving and accumulating dynamic of freshwater systems can enable the propagation of anthropogenic disturbances over long distances.

We recommend that future efforts focus on enhancing the quality of existing sets of geospatial data and on providing more reliable data indicating the presence,

Table 7.4 Data and conceptual gaps that typically hinder evaluations of the biodiversity value and ecological integrity of aquatic systems, as well as impede the eventual development of effective conservation strategies. The specific nature of these gaps tends to vary widely across systems

Data gaps
Species inventories and distribution maps, and meta-population structures
Species habitat requirement information
High resolution maps of streams, riparian zones, floodplains, and wetlands
Characteristics of natural flow regimes, derived from long-term gauging data
Ground-truthed drainage maps
Spatially explicit maps of runoff generation
Data describing groundwater-surface water relationships
Degree of imperilment of individual species or populations, and causes of imperilment
Invasive exotic species maps
Geo-referenced data on dams, roads, and other infrastructure
Geo-referenced data on water withdrawals and returns
Inland fisheries data
Conceptual gaps
How do different human activities exert a greater/lesser influence over a variety of spatial and temporal scales?
How do different human activities interact to affect ecological integrity?
How does the configuration of land uses, including the size of blocks and the proximity to aquatic systems, affect ecological integrity?
What confers resilience to freshwater systems, and how can we maximize it?
How can we determine thresholds in land use and other parameters that, when exceeded, cause unacceptable declines in ecological integrity?
How can human needs for water and other freshwater resources be met while maintaining native freshwater biodiversity?

location, and conditions of salient hydrologic features (e.g., geographically referenced dams and their operational schemes, biogeochemical conditions of lakes, altered and pristine river courses, and the quantity and quality of surface and groundwater). Likewise, analyses would benefit from improvements in species and habitat distribution (see gaps, Table 7.4).

We strongly recommend that freshwater ecologists pursue studies that seek to explain the relationships between common land uses and the quality of affected freshwater habitats. These studies would be most helpful to conservation managers and planners if they focus on the scales over which land use-related disturbances operate and discrete thresholds above which ecologically meaningful changes occur – whether these changes occur in the biotic (e.g., species assemblages), abiotic (e.g., habitat features), or hydrologic realms (see Table 7.4). The ongoing development of indices of ecological integrity could both benefit from, and help focus, such research. However, at current levels of investment in this kind of research, it seems difficult to hope for the amount of meaningful progress in aquatic biodiversity conservation and management that could be expected to stem the incoming tide of aquatic extinctions. To reduce the potential magnitude of aquatic biodiversity loss, governments worldwide will need to promote long-term national visions focused on the many benefits of sustaining their nation's patrimony of freshwater ecosystems and the remaining native species that are those systems' inhabitants.

References

Abell RA, Olson DM, Dinerstein E, Hurley PT, Diggs JT, Eichbaum W, Walters S, Wettengel W, Allnutt T, Loucks CJ, Hedao P (2000) Freshwater ecoregions of North America: a conservation assessment. Island Press, Washington, DC

Abell R, Thieme M, Dinerstein E, Olson D (2002) A sourcebook for conducting biological assessments and developing biodiversity visions for ecoregion conservation. Volume II: Freshwater ecoregions. World Wildlife Fund-US, Washington, DC

Abell R, Thieme M, Revenga C, Bryer M, Kottelat M, Bogutskaya N, Coad B, Mandrak N, Contreras-Balderas S, Bussing W, Stiassny MLJ, Skelton P, Allen GR, Unmack P, Naseka A, Ng R, Sindorf N, Robertson J, Armijo E, Higgins J, Heibel TJ, Wikramanayake E, Olson D, Lopez HL, Reis RE, Lundberg JG, Sabaj Perez MH, Petry P (2008) Freshwater ecoregions of the world: a new map of biogeographic units for freshwater biodiversity conservation. Bioscience 58:403–414

Alcamo J, Döll P, Henrichs T, Kaspar F, Lehner B, Rösch T, Siebert S (2003) Development and testing of the water GAP 2 global model of water use and availability. Hydrol Sci J 48(3): 317–337

Allan JD, Flecker AS (1993) Biodiversity conservation in running waters: identifying the major factors that threaten destruction of riverine species and ecosystems. Bioscience 43:32–43

Allan JD, Abell R, Hogan Z, Revenga C, Taylor BW, Welcomme RL, Winemiller K (2005) Overfishing of inland waters. Bioscience 55(12):1041–1051

Berkes F, Colding J, Folke C (2000) Rediscovery of traditional ecological knowledge as adaptive management. Ecol Appl 10:1251–1262

Boon PJ (2000) The development of integrated methods for assessing river conservation value. Hydrobiologia 422(423):413–428

Boulton AJ (1999) An overview of river health assessment: philosophies practice problems and prognosis. Freshw Biol 41:469–479

Caraco NF, Cole JJ, Likens GE, Lovett GM, Weathers KC (2003) Variation in NO_3 export from flowing waters of vastly different sizes: does one model fit all? Ecosystems 6:344–352

Center for International Earth Science Information Network, Centro Internacional de Agricultura Tropical (2005) Gridded population of the world, version 3 (GPWv3): Population Grids. Socioeconomic Data and Applications Center (SEDAC) Columbia University, Palisades NY. Available at http://sedac.ciesin.columbia.edu/gpw

Chao NL, Prang G (1997) Project Piaba: towards a sustainable ornamental fishery in the Amazon. Aquarium Sci Conserv 1:105–111

Collen B, Ram M, Dewhurst N, Clausnitzer V, Kalkman VJ, Cumberlidge N, Baillie JEM (2008) Broadening the coverage of biodiversity assessments. In: Vié J-C, Hilton-Taylor C, Stuart SN (eds) Wildlife in a changing world: an analysis of the 2008 IUCN red list of threatened species. IUCN, Gland, Switzerland, pp 67–76

Cowx IG (2002) Analysis of threats to freshwater fish conservation: past and present challenges. In: Collares-Pereira MJ, Cowx IG, Coelho M (eds) Conservation of freshwater fishes: options for the future. Fishing News Books, Oxford, pp 201–220

Darwall W, Smith K, Lowe T, Vié J-C (2005) The status and distribution of freshwater biodiversity in Eastern Africa, vol 31, Species survival commission occasional paper. World Conservation Union, Gland

De Leo GA, Levin S (1997) The multifaceted aspects of ecosystem integrity. Conserv Ecol 1: 3. Available at http://www.consecol.org/vol1/iss1/art3

Döll P, Kaspar F, Lehner B (2003) A global hydrological model for deriving water availability indicators: model tuning and validation. J Hydrol 270:105–134

Dudgeon D, Arthington AH, Gessner MO, Kawabata ZI, Knowler DJ, Leveque C, Naiman RJ, Prieur-Richard AH, Soto D, Stiassny MLJ, Sullivan CA (2006) Freshwater biodiversity: importance threats status and conservation challenges. Biol Rev 81:163–182

Dynesius M, Nilsson C (1994) Fragmentation and flow regulation of river systems in the northern third of the world. Science 266:753–762

Environmental Systems Research Institute (1993) Digital Chart of the World. Environmental Systems Research Institute, Redlands, Calif. Available at http://ortelius.maproom.psu.edu/dcw/

Environmental Systems Research Institute (1992) ArcWorld 1:3M. Environmental Systems Research Institute, Redlands, Calif

FAO (1998) Database on the introduction of aquatic species (DIAS). Food and Agriculture Organization of the United Nations, Rome. Available at http://www.fao.org/fi/statist/fisoft/dias/index.htm

FAO (1999) Review of the state of world fishery resources: Inland fisheries, vol 942, FAO fisheries circular. Food and Agriculture Organization of the United Nations, Rome

FAO (2001) World river sediment yields database. FAO, Rome. Available at http://www.fao.org/landandwater/aglw/sediment/default.asp

FAO (2003) Africover. Food and Agriculture Organization of the United Nations, Rome. Available at http://www.africover.org/

Farr TG, Kobrick M (2000) Shuttle radar topography mission produces a wealth of data. Eos Trans Am Geophys Union 81:583–585

Gergel SE, Turner MG, Miller JM, Melack JM, Stanley EH (2002a) Landscape indicators of human impacts to riverine systems. Aquat Sci 64:118–128

Gergel SE, Dixon MD, Turner MG (2002b) Consequences of human-altered floods: levees floods and floodplain forests along the Wisconsin River. Ecol Appl 12:1755–1770

Gleick PH (2000) The changing water paradigm: a look at twenty-first century water resources development. Water Int 25:127–138

Global International Water Assessment (2002) Water issues: causal chain analysis. Global International Water Assessment, Kalmar, Sweden. Available at http://www.giwa.net/caus_iss/causual_chain_analyses.phtml

Global Water System Project (2005) The Global Water System Project: science framework and implementation activities. Global Water System Project, Bonn. Available at http://www.gwsp.org
Gregory SV, Swanson FJ, McKee WA, Cummins KW (1991) An ecosystem perspective of riparian zones. Bioscience 41:540–551
Groves CR (2003) Drafting a conservation blueprint: a practitioner's guide to planning for biodiversity. Island Press, Washington, DC
Groves CR, Jensen DB, Valutis LL, Redford KH, Shaffer ML, Scott JM, Baumgartner JV, Higgins JV, Beck MW, Anderson MG (2002) Planning for biodiversity conservation: putting conservation science into practice. Bioscience 52:499–512
Harding JS, Benfield EF, Bolstad PV, Helfman GS, Jones EBD (1998) Stream biodiversity: the ghost of land use past. Proc Natl Acad Sci USA 95:14843–14847
Harrison IJ, Stiassny MLJ (1999) The quiet crisis: a preliminary listing of the freshwater fishes of the world that are extinct or 'missing in action'. In: MacPhee RDE (ed) Extinctions in near time: causes contexts and consequences. Kluwer Academic, New York, pp 271–332
Hearn P, Hare T, Schruben P, Sherrill D, Lamar C, Tsushima P (2000) Global GIS Database. US Geological Survey, Flagstaff. Available at http://webgis.wr.usgs.gov/globalgis/
Hodges JCF, Friedl MA, Strahler AH (2001) The MODIS global land cover product: new data sets for global land surface parameterization. Proc Global Change Open Sci Conf Amsterdam. Available at http://geography.bu.edu/landcover/userguidelc/
Hughes RM, Hunsaker CT (2002) Effects of landscape change on aquatic biodiversity and biointegrity. In: Gutzwiller KJ (ed) Applying landscape ecology in biological conservation. Springer, Berlin, pp 309–329
International Commission on Large Dams (2007) World register of Dams. International Commission on Large Dams, Paris. Available at www.icold-cigb.net
IUCN (2001) Keeping on eye on threatened species: the IUCN Red List. World Conserv 3.
IUCN (2008) IUCN Red list of threatened species. World Conservation Union, Gland. Available at www.iucnredlist.org
Joint Research Centre, European Commission (2003) Global Land Cover (2000) database. Joint Research Centre European Commission, Ispra, Italy. Available at http://gem.jrc.ec.europa.eu/products/glc(2000)/glc(2000).php
Jones JA, Swanson FJ, Wemple BC, Snyder KU (2000) Effects of roads on hydrology geomorphology and disturbance patches in stream networks. Conserv Biol 14:76–85
Karr JR, Dudley DR (1981) Ecological perspective on water quality goals. Environ Manage 5:55–68
Lehner B, Döll P (2004) Development and validation of a global database of lakes reservoirs and wetlands. J Hydrol 296:1–22
Lehner B, Verdin K, Jarvis A (2008) New global hydrography derived from spaceborne elevation data. Eos Transact 89(10):93–94. Available at http://hydrosheds.cr.usgs.gov/
Lehner B, Reidy Liermann C, Revenga C, Fekete B, Vörösmarty C, Crouzet P, Döll P, Endejan M, Frenken K, Magome J, Nilsson C, Robertson JC, Rödel R, Sindorf N, Wisser D (in review): High resolution mapping of the world's reservoirs and dams for sustainable river flow management. Submitted to Frontiers in Ecology and the Environment, http://www.gwsp.org/current_activities.html
Ligon FK, Dietrich WE, Trush WJ (1995) Downstream ecological effects of dams. Bioscience 45:183–192
Malmqvist B, Rundle S (2002) Threats to the running water ecosystems of the world. Environ Conserv 29:134–153
McAllister DE, Parker BJ, McKee PM (1985) Rare, endangered, and extinct fishes in Canada, vol 54, Syllogeus. National Museums of Natural Sciences National Museums of Canada, Ottawa
McAllister DE, Hamilton AL, Harvey B (1997) Global freshwater biodiversity: striving for the integrity of freshwater ecosystems. Sea Wind 11:1–140
Miller RR, Williams JD, Williams JE (1989) Extinctions of North American fishes during the past century. Fisheries 14:22–38

Niger Basin Authority (ABN) (2007) Niger river basin atlas. ABN, Niamey, Niger
Nilsson C, Svedmark M, Hansson P, Xiong S, Berggren K (2000) River fragmentation and flow regulation analysis. unpublished data. Umeå University, Umeå, Sweden.
Nilsson C, Reidy CA, Dynesius M, Revenga C (2005) Fragmentation and flow regulation of the world's large river systems. Science 308:405–408
Norris RH, Thoms MC (1999) What is river health? Freshw Biol 41:197–209
Noss RF (1992) The Wildlands Project: land conservation strategy. Wild Earth (Suppl, The Wildlands Project):10–25
O'Keeffe JH, Danilewitz DB, Bradshaw JA (1987) An 'expert system' approach to the assessment of the conservation status of rivers. Biol Conserv 40:69–84
O'Neill RV, Hunsaker CT, Jones KB, Riitters KH, Wickham JD, Schwartz PM, Goodman IA, Jackson BL, Baillargeon WS (1997) Monitoring environmental quality at the landscape scale: using landscape indicators to assess biotic diversity watershed integrity and landscape stability. Bioscience 47:513–519
Oak Ridge National Laboratory (2009) LandScan global population database. Oak Ridge National Laboratory, Oak Ridge. Available at http://www.ornl.gov/landscan/
Peterson BJ, Wollheim WM, Mulholland PJ, Webster JR, Meyer JL, Tank JL, Martí E, Bowden WB, Valett HM, Hershey AE, McDowell WH, Dodds WK, Hamilton SK, Gregory S, Morrall DD (2001) Control of nitrogen export from watersheds by headwater streams. Science 292:86–90
Poff NL, Hart DD (2002) How dams vary and why it matters for the emerging science of dam removal. Bioscience 52:659–668
Poff NL, Brinson MM, Day JW (2002) Aquatic ecosystems and global climate change: potential impacts on inland freshwater and coastal wetland ecosystems in the United States. Pew Center on Global Climate Change, Arlington
Postel S (2002) Rivers of life: the challenge of restoring health to freshwater ecosystems. Water Sci Technol 45:3–8
Reaka-Kudla M (1997) The global biodiversity of coral reefs: a comparison with rain forests. In: Reaka-Kudla ML, Wilson DE, Wilson EO (eds) Biodiversity II: understanding and protecting our biological resources. Joseph Henry, Washington, DC, pp 83–103
Revenga C, Brunner J, Henninger N, Kassem K, Payne R (2000) Pilot analysis of global ecosystems: freshwater systems. World Resources Institute, Washington, DC
Ricciardi A, Rasmussen JB (1999) Extinction rates of North American freshwater fauna. Conserv Biol 13:1220–1222
Rosenberg DM, McCully P, Pringle CM (2000) Global-scale environmental effects of hydrological alterations. Bioscience 50(9):746–751
Saunders DL, Meeuwig JJ, Vincent ACJ (2002) Freshwater protected areas: strategies for conservation. Conserv Biol 16:30–41
Sedell J, Sharpe M, Apple D, Copenhagen M, Furniss M (2000) Water and the forest service. FS-660. USDA Forest Service, Washington, DC
Sidle RC, Sasaki S, Otsuki M, Noguchi S, Rahim Nik A (2004) Sediment pathways in a tropical forest: effects of logging roads and skid trails. Hydrol Process 18:703–720
Siebert S, Döll P, Hoogeveen J, Faures J-M, Frenken K, Feick S (2005) Development and validation of the global map of irrigation areas. Hydrol Earth Syst Sci 9:535–547
Smakhtin V, Revenga C, Döll P (2004) A pilot global assessment of environmental water requirements and scarcity. Water Int 29(3):307–317
Smith KG, Darwall WRT (2006) The status and distribution of freshwater fish endemic to the Mediterranean Basin. IUCN, Gland
Stein BA, Kutner LS, Adams JS (eds) (2000) Precious heritage: the status of biodiversity in the United States. Oxford University Press, New York
Stuart SN, Chanson JS, Cox NA, Young BE, Rodrigues ASL, Fischman DL, Waller RW (2004) Status and trends of amphibian declines and extinctions worldwide. Science 306:1783–1786

Thieme ML, Abell RA, Stiassny MLJ, Skelton PH, Lehner B, Teugels GG, Dinerstein E, Kamdem-Toham A, Olson DM (2005) Freshwater ecoregions of Africa: a conservation assessment. World Wildlife Fund-US, Washington, DC

Tockner K, Stanford JA (2002) Riverine flood plains: present state and future trends. Environ Conserv 29:308–330

Turner RE, Rabalais NN, Justic D, Dortch Q (2003) Global patterns of dissolved N, P and Si in large rivers. Biogeochemistry 64:297–317

Turtle Conservation Fund (2002) Global action plan for conservation of tortoises and freshwater turtles, strategy and funding prospectus, 2002–2007. Conservation International and Chelonian Research Foundation, Washington, DC

United Nations (1997) Comprehensive assessment of the freshwater resources of the world. World Meteorological Organization and Stockholm Environment Institute, Geneva

United Nations Environment Programme/Global Resource Information Database, Center for International Earth Science Information Network, World Resources Institute, National Center for Geographic Information and Analysis (2004) African population database, version 4. United Nations Environment Programme/Global Resource Information Database, Sioux Falls. Available at http://na.unep.net/globalpop/africa/

US Army Corps of Engineers (2007) National inventory of dams. United States Army Corps of Engineers, Alexandria. Available at https://nid.usace.army.mil

US Geological Survey, University of Nebraska-Lincoln, Joint Research Centre of the European Commission (1997) Global land cover characterization (GLCC), version 1.2. USGS, Washington, DC. Available at http://edcdaac.usgs.gov/glcc/glcc.html

US National Oceanographic and Atmospheric Administration/National Geophysical Data Center (1998) Stable lights and radiance calibrated lights of the world CD-ROM. National Oceanic and Atmospheric Administration-National Geophysical Data Center, Boulder. Available at http://spidr.ngdc.noaa.gov/

Vörösmarty CJ, Sharma KP, Fekete BM, Copeland AH, Holden J, Marble J, Lough JA (1997) The storage and aging of continental runoff in large reservoir systems of the world. Ambio 26:210–219

Vörösmarty CJ, McIntyre PB, Gessner MO, Dudgeon D, Prusevich A, Green P, Glidden S, Bunn SE, Sullivan CA, Reidy Liermann C, Davies PM (2010) Global threats to human water security and river biodiversity. Nature 467:555–561

Wang L, Lyons J, Kanehl P, Gatti R (1997) Influences of watershed land use on habitat quality and biotic integrity in Wisconsin streams. Fisheries 22:6–12

Welcomme RL (2001) Inland fisheries: ecology and management. Blackwell Science, Rome

Welcomme R (2005) Impacts of fishing on inland fish populations in Africa. In: Thieme ML, Abell RA, Stiassny MLJ, Skelton PH, Lehner B, Teugels GG, Dinerstein E, Kamdem-Toham A, Olson DM (eds) Freshwater ecoregions of Africa: a conservation assessment. World Wildlife Fund-US, Washington, DC

Wetlands International (2002) Special issue on the Niger Basin Initiative. Fadama Newsletter of the Wetlands International Africa Office, 5. Dakar-Yoff, Senegal

Wetlands International (2007) Ramsar Sites Database. Wetlands International, Wageningen. Available at http://ramsar.wetlands.org/Default.aspx

Williams JD, Warren ML, Cummings KS, Harris JL, Neves RJ (1993) Conservation status of freshwater mussels of the United States and Canada. Fisheries 18:6–22

World Commission on Dams (2000) Dams and development. Earthscan, London

World Resources Institute, WCMC World Conservation Monitoring Centre, PADCO, Inc (1995) Africa data sampler. CD-ROM. World Resources Institute, Washington, DC

Part II
Interactions in Specific Ecosystems

Chapter 8
A Cross-Cultural Analysis of Human Impacts on the Rainforest Environment in Ecuador

Flora Lu and Richard E. Bilsborrow

8.1 Introduction

Perhaps no biome on the planet has higher biodiversity than the Amazon rainforest, which covers a mere 7% of the landmass but contains an estimated half of all species. Although there are many ecoregions within the Amazon basin noteworthy for their ecological richness, several have particularly high species diversity, perhaps the most notable being the eastern slopes of the Andes of southern Colombia, Ecuador, and Peru. Here, nutrient-rich sediments originating in the Andes, topographic variation, tropical climate, and high levels of rainfall converge to harbor a wealth of biodiversity and endemism. According to the tropical ecologist Norman Myers (1988), western Amazonia "is surely the richest biotic zone on Earth, and deserves to rank as a kind of *global epicenter of biodiversity*" (italics added). For instance, the Ecuadorian Amazon or *Oriente* houses an estimated 9,000–12,000 species of vascular plants. In Ecaudor's 600,000-acre Yasuni National Park, a United Nations Educational, Scientific, and Cultural Organization (UNESCO) World Biosphere Reserve, some years ago scientists identified more than 600 species of birds, 500 species of fish, and 120 species of mammals (cited in Kimerling 1991: 33). Moreover, a detailed assessment of tree biodiversity in 16 tropical sites around the world conducted by the Smithsonian Tropical Research Institute concluded that Yasuni contained 1,104 species (of at least one cm dbh) in a 25 ha area, the most of any site studied (Romoleroux et al. 1997; Pitman et al. 2002). In the Cuyabeno Reserve in northeastern Ecuador – another important Amazonian conservation area within our study region – 313 species of trees were identified within a single hectare, and 500 species of birds and 100 species of mammals (http://www.worldwildlife.org/wildworld/profiles/terrestrial_nt.html) have also been reported (see also Valencia et al. 1994). In this chapter, we focus our attention on this ecologically valuable and vulnerable region, examining a cross-cultural sample of indigenous peoples who are at the center of a vortex of changing economic, ecological, and cultural dynamics.

Confronting the ecological complexity of the region are sociocultural, economic, and political processes that have produced rapid deforestation and land cover change over the past three decades, with Ecuador having the highest rate of

R.P. Cincotta and L.J. Gorenflo (eds.), *Human Population: Its Influences on Biological Diversity*, Ecological Studies 214,
DOI 10.1007/978-3-642-16707-2_8,

deforestation in the 1990s of the seven countries constituting the Amazon basin (Food and Agriculture Organization of the United Nations 2005). The complexity of these processes defies simple explanation as many factors appear to be involved at different scales, including road construction, petroleum exploration, and the expansion of agriculture by colonists (e.g., Rudel 1983; Pichón 1997; Pan and Bilsborrow 2005). The changing relationships between indigenous peoples and their environments have been viewed in diametrically opposite ways, with indigenous populations alternately portrayed as a solution and as a major threat to conservation. In the former portrayal, they are depicted as static, isolated, ecologically noble conservationists living in harmony with nature as they presumably have for centuries (Brosius 1997; Conklin and Graham 1995; Redford 1990). More recently, however, some biologists have become disillusioned due to the contrast between the real-world behaviors of indigenous peoples and the "ecologically noble savage" ideal, perceiving indigenous resource use as yet another threat to ecological viability (e.g., Terborgh 1999). The common perception of Native Amazonians in these debates as homogeneous in terms of cultural values and use of land and resources is a great oversimplification, with negative ramifications for developing nuanced policies and long-lasting collaborations between indigenous and nonindigenous stakeholders. Conklin and Graham write, "Representations of Amazonian Indians circulating in the international public sphere tend to be generic stereotypes that misrepresent the diversity of native Amazonian cultures and the complexity of native priorities and leadership issues" (1995: 705). An examination of indigenous land use patterns should therefore take into account cultural, economic, and demographic variation. As indigenous groups become more integrated into market economies (e.g., Fisher 2000; Godoy 2001), experience accelerated sociocultural change, adopt more sedentary settlement patterns, acquire titles to huge areas of tropical forests, and yet at the same time find their lands increasingly circumscribed by alternative land users and land use, efforts to understand the changing human-environment relationships of indigenous populations become of paramount importance.

In this chapter, we provide a multifaceted examination of indigenous lifeways and resource use of five ethnic groups in the northern Ecuadorian Amazon region. Using ethnographic data collected from eight communities, and survey data from 36 communities encompassing five different ethnicities, we compare and contrast the indigenous populations in terms of demographic characteristics, involvement in the market economy, and patterns of forest conversion and faunal exploitation. This analysis helps put to rest ideas of generic stereotypes of Native Amazonian populations and highlights the complexities in human–environment relationships, which lie at the heart of both conservation and cultural survival.

8.2 The Ecuador Research Project

In 2000, the authors initiated a research project investigating cross-cultural patterns of indigenous land use in the northern Ecuadorian Amazon. The objective was to determine the demographic, socioeconomic, and biophysical factors influencing land and resource use of five indigenous populations – the Quichua, Shuar, Cofán, Secoya, and Huaorani. This requires a multidisciplinary and collaborative approach that combines quantitative and qualitative methodologies from demography, geography, and anthropology. Data collection, carried out in 2001, involved two phases of research: (1) an ethnographic study in eight indigenous communities and (2) household and community surveys in 28 additional communities. In addition, Global Positioning System (GPS) receivers were used in the field to obtain geographic coordinates of roads, communities, households and their agricultural plots, and important infrastructure, such as schools and health clinics. Satellite imagery was processed to determine land cover types, land use, landscape features, and the location of roads and other key infrastructure. The georeferenced socioeconomic and demographic household survey data and biophysical and remotely sensed data have been integrated in a geographic information system (GIS) to derive measures of land cover and to use for multivariate analyses and spatial analysis at levels from the household farm plot of land to the landscape and the region.

The first phase of data collection was an ethnographic study carried out from February to June, 2001 (Holt et al. 2004). Ethnographic researchers were placed in eight indigenous communities for a 5-month period (see data in Table 8.1 and locations in Fig. 8.1). Given the much larger size of the Quichua population, the purposively selected sample included two Quichua villages, in addition to one village each for the Shuar, Secoya, Cofán, and Huaorani. But two of the sample villages had recently splintered, resulting in a final ethnographic sample of eight villages, including three Quichua and two Huaorani villages. The total number of households studied was 120, comprising 677 individuals. Ethnographic data collection included participant observation, time allocation, household economic diaries, and formal questionnaires covering demographic behavior and attitudes, agricultural production and resource use, household economics, and socioeconomic attitudes.

Hunting data were gathered in 2001 through posthunt interviews and formal questionnaires. A posthunt data form was developed which ascertained hunter, ethnicity, date, time departed and returned, tools used, and number of other participants in the hunt (people and dogs). Then, for each animal encountered, investigators

Table 8.1 Communities included in ethnographic study

Community	Ethnicity	No. of households	% Households	No. of persons	Mean household size
Pastaza Central	Quichua	10	8.3	57	5.7
Pachacutik	Quichua	11	9.2	79	7.2
Sewaya	Secoya	20	16.7	97	4.9
Zábalo	Cofán	27	22.5	133	4.9
El Pilche	Quichua	22	18.3	131	6.0
Tiguano	Shuar	13	10.8	70	5.4
Quehueiri-ono	Huaorani	10	8.3	67	6.7
Huentaro	Huaorani	7	5.8	43	6.1
Total		120	100.0	677	5.6

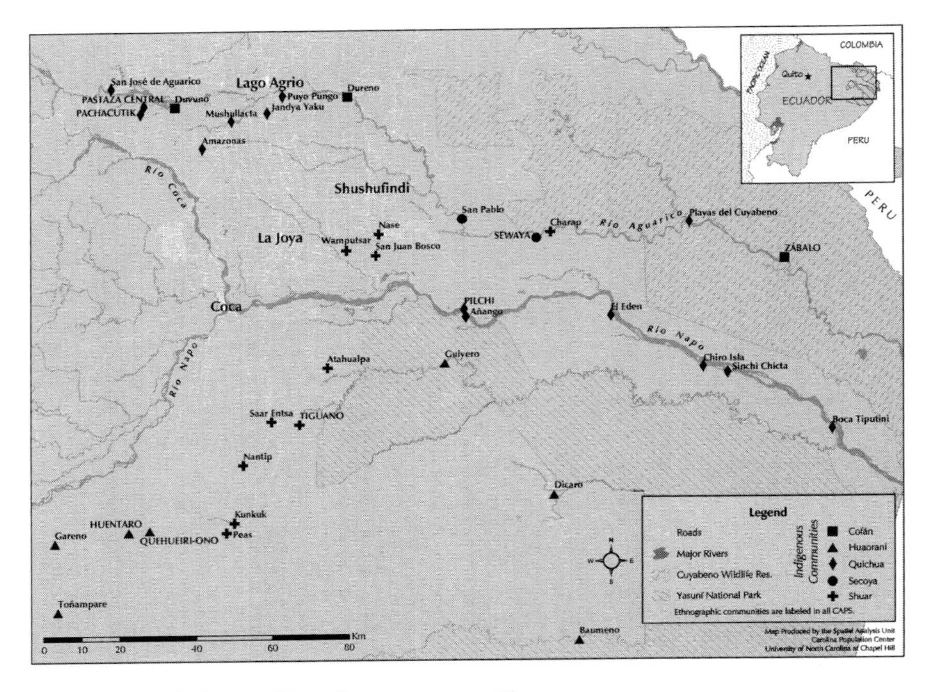

Fig. 8.1 Map of ethnographic and survey communities

asked the Spanish and indigenous name; type of animal (bird, mammal, etc.); distance from the community where encountered; whether it was pursued and for how long; if it was killed, its age and sex, and the procurement technology used. Equivalent information was also collected for animals killed using traps, which are employed mainly in the vegetable garden or agricultural plots called *chacras*. In total, 364 post-hunt interviews were carried out in the eight indigenous communities, involving 1,148 animal encounters. For 1,031 of those animal encounters, data are available on whether the animal was killed or not. In 679 of those cases, the animal was killed, providing reliable information for analysis. In addition, in the formal household survey questionnaires, over 500 men in 36 communities were asked about the following: hunting participants (number, sex, and age); tools used; choice of prey and perceived abundance; frequency and success of hunting; whether sold meat or live animals; location/distance of hunts; frequency of night hunting; and use of dogs. A shortened version of the male hunting questionnaire was administered to women to ascertain their participation in hunting, including frequency, prey caught or killed, location, procurement technology, and yield.

Different labor tasks and their variation by age, gender, household size/composition, and ethnicity were studied using the spot check time allocation method (Borgerhoff-Mulder and Caro 1985). The time allocation data were collected through randomized household visits. Using a table of random times between 6:00 am and 7:00 pm, ethnographers visited all logistically feasible households in the community using a circuit whose starting point was not fixed. The following data were noted on each data form: name of the community, name of the head of household, date of observation, observer, and time of observation. For each member of the household, researchers noted the subject's name, activity code(s), and location. The reliability of information was also noted: whether the researcher observed the subject first or vice versa, whether the subject was absent during the time of observation and data were provided by a proxy respondent, and whether any visitors to the household were present. While there was a total study population of 677 in the eight communities studied in 2001, time allocation observations were obtained only for individuals aged 5 years or more, reducing the sample actually used to 509 people.[1] There were a total of 5,694 household visits, generating 23,796 person-observations during the 5 month study period.

In order to collect quantitative data about the household economy, an "Input/Output Household Diary" data sheet (Diario de Ingresos/Egresos) was developed to record systematically the daily flow of goods and services into and out of each household. To this end, households were asked to keep detailed daily diaries of the source (person, institution) and quantity of goods or money coming into the household, as well as any "expenditures" in money or goods leaving the household. Inputs included: game, fish, or plant (including agricultural) items collected; monetary income from the sale of crops, game, domestic animals, or handicrafts; items received in exchange for labor; and gifts from other households or outsiders. Outputs included: cash outlays for the purchase of food and drink, household items, personal items, medicine, and agricultural inputs; materials or foods given to another household; payment (in food or money) to others for their labor; and money spent on travel or recreation.

We also collected rough data on dietary intake, using a short checklist for noting items consumed by the household that day in general categories, without reference to quantities. The food categories included in this dietary checklist were: forest game, domestic or purchased meat, fish, dairy, eggs, insects, legumes, grains and manioc, fruits, vegetables, nuts, and seeds. A person in each household interested in filling in the diary was trained by ethnographers on how to complete the form. The ethnographers collected completed forms and dropped off blank ones every few days and checked the data for consistency and reliability. In total, for the eight communities, 89 of the 120 households (74%) participated in the household economic diaries, and 4,041 household-days of "input–output" sheets were completed.

The sections below on demography and land use draw from the *second* phase of data collection a household and community survey implemented in 36 indigenous communities (the eight ethnographic communities plus 28 additional communities – see Gray et al. 2008, for sampling procedures) Fig. 8.2. Interviews were conducted separately with male and female household heads (or the head and spouse) by male

---

[1]In some communities, several households far from the community center were excluded from the time use study and the household diaries as they could not be routinely visited in the circuit of random visits.

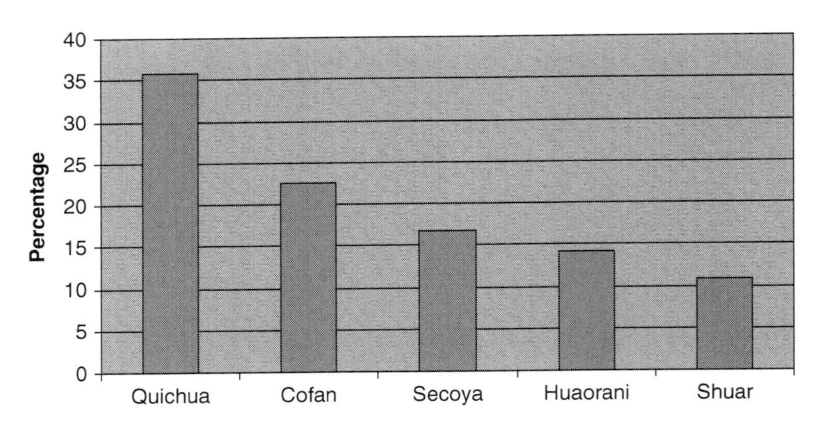

Fig. 8.2 Distribution of households in survey by ethnic group (2001)

and female interviewers, respectively, usually in Spanish.[2] The head's questionnaire covered migration history, land tenure and use, production and sale of crops and cattle, wage employment in the community and elsewhere, hunting and fishing, foraging, and technical assistance and credit received, among other topics. Questions on land use elicited information on the size, composition, age, and distance from the dwelling of all plots currently in use. Intercropped areas were divided into their constituent land uses based on proportional coverage, and the names of all crops occupying 10 m^2 or more were recorded. The spouse's questionnaire included a household roster and asked about her own migration history, the out-migration of household members and household assets, and the roles of members in decision making, fertility, mortality, and illnesses and health care, among other topics. Finally, a community-level survey collected data from community leaders on population size, community infrastructure and organization, location relative to roads and major towns, means and frequency of transport, and contact with external institutions.

8.3 Cultural Diversity in Ecuador's Amazon

The five indigenous populations studied vary greatly in terms of population density, integration with the market economy, access to land and resources, and cultural values. Below we first provide ethnographic sketches of each ethnic group, followed by basic comparative demographic data on population size, sex–age distribution, fertility, mortality, migration, and population growth. This section concludes with a

[2]When necessary, an interpreter was recruited from the community to assist. This was necessary only about 10% of the time, mainly with older women in Cofán and Huaorani communities.

discussion of the varying sources of livelihood in each community and their degree of integration with the market economy.

8.3.1 Ethnographic Sketches

The Shuar are the members of the Jivaroan language group concentrated near the Peru/Ecuadorian border. Numbering about 40,000 persons, they are the second largest indigenous population in the Ecuadorian Amazon. The Shuar's history of sustained contact with outsiders began with Catholic priests in the early twentieth century. Partly as an attempt to protect their lands against colonist incursions, they have adopted cattle production, which requires clearing large areas of forest, to secure land claims (Hendricks 1988). The Shuar have also reorganized themselves from living in dispersed households to forming nucleated *centros* or communities on 3,000–6,000-ha tracts (Rudel et al. 2002). Originally inhabiting the southern Amazonian province of Morona Santiago, many Shuar migrated north to the four northern Amazonian provinces of Pastaza, Orellana, Napo, and Sucumbios in search of land. Our study includes such a migrant Shuar community, Tiguano, in Orellana province – at the time of the study in 2001, about 3 hours by vehicle south of Coca. This group is not necessarily representative of the larger Shuar population, but is similar to other Shuar population groups in the study region.

The Secoya live along the Aguarico River and its tributaries downriver from Lago Agrio and belong to the Western Tocanoan linguistic family. Once estimated at 12,000, their population drastically declined during the Spanish conquest due to sickness and slavery. They currently number only about 700 people in Ecuador and Peru combined (Cabodevilla 1989, 1997; Vickers 1989). They live in scattered households or small villages along the banks of rivers (mainly the Aguarico River in Ecuador) and streams. Traditionally, they relocated their settlements every 5–20 years. In 1996, the Secoya territory was legalized as "Centro Siecoya Remolino", which encompassed 23,000 ha of land, including part on the northern bank of the Aguarico in the ecologically rich Cuyabeno Wildlife Refuge. In 2001, their territory was officially increased by another 2,807 ha in the Cuyabeno reserve (De la Torre et al. 2000).

The Quichua (also spelled Kichwa), or *Runa*, are the most numerous of Ecuador's Native Amazonian peoples, with an estimated 60,000 people in Sucumbios, Orellana, Napo, and Pastaza provinces (Irvine 2000). Like the Shuar, they have a long history of contact with outsiders; indeed, the lowland Amazonian Quichua first emerged as a distinct ethnic group when preexisting indigenous societies were decimated by disease, violence, and social disruption during the Spanish conquest. Under the violence and repression suffered at the hands of the Spanish, survivors from these different ethnic groups decided to or were obliged to live in mission villages where Quichua, an Andean language, served as a *lingua franca*. A shared Quichua ethnic identity is thought to have emerged around 1800. Today, the Runa are divided into three distinct cultural and linguistic subgroups. Pastaza Province (southern part of their territory) is home to the Canelos Runa, while the higher elevation (600 m and above) northern region is inhabited by the Napo Runa (clustered around the towns of Archidona and Tena), while the lower elevations are home to the Loreto Runa (centered around the towns of Loreto, Avila, and San José de Payamino). Boundaries between these groups, once distinct, have become blurred over time (Irvine 2000: 24–25).

The A'i people, or Cofán, traditionally occupied the area between the San Miguel River, the Guamuez River, the Bajo Putumayo River, and the Aguarico in southern Colombia and northern Ecuador. Their origin is unknown, some believing the A'i language to be unique, while others think it belongs to the linguistic Chibcha family of Colombia (Califano and Gonzalo 1995; Cerón 1995). Negatively impacted by outsiders at the end of the nineteenth century during the rubber and quinine booms, the Cofán again suffered beginning about 1970 when they were displaced by petroleum extraction. They were forced to move from the region around Lago Agrio (where significant oil deposits were first discovered in the Ecuadorian Amazon in 1967) and formed scattered settlements, mostly further to the east, deeper in the forest. Currently, the Cofán of Ecuador number approximately 500 persons in five communities. Through an agreement with the Ministerio de Agricultura y Ganadería, the Cofán were given legal title to 80,000 ha in 1992. Nine years later, in 2001, this area was expanded by 50,000 ha around the Guepi River (Albuja et al. 2000).

The Huaorani (also spelled Waorani) were contacted peacefully for the first time by Protestant missionaries in 1958, the last of the indigenous groups in the Ecuadorian Amazon to be contacted by outsiders. Indeed, some Huaorani subgroups still have not been peacefully contacted, and they fiercely resist what they consider to be intrusions by outsiders. From a population numbering only about 500 at the time of missionary contact, the number of Huaorani is growing, estimated at about 2,000 persons (Lu 1999). Their language, *huao tededo*, is a linguistic isolate, and their reputation for warfare and spearing raids allowed them to occupy and claim a large territory bordered on the north by the Napo River and on the south by the Curaray River (approximately 20,000 km^2) before sustained contact. They currently have legal title to about one-third of that area. Before missionary contact, their settlement pattern was characterized by dispersed and autonomous *nanicaboiri* (longhouses) of closely related kin; now small, nucleated communities are more common, centered around a school and sometimes a landing strip.

8.3.2 Demography

Table 8.1 indicates the population size of each of the eight indigenous ethnographic study communities, by ethnicity and mean household size. Mean household size is highest in the Quichua community of Pachacutik at 7.2 persons and lowest for the Cofán community of Zábalo and the Secoya community of Sewaya at 4.9 persons per household. In all eight communities, males constitute a majority, ranging from

51% to 60% of the residents. The populations are all very young, with half or more under age 20, reflecting the high fertility of women in the villages. In fact, fertility is so high that even with the prevailing moderate levels of mortality, natural population growth is high, on average around 3% per year.

Data from the survey phase of the project indicate not only high overall levels of fertility, as expected, but also wide variations across ethnicities and communities, which are not what most observers would expect. As the sample of women in the ethnographic study is far too small for reliable estimates, we draw upon data from the larger household survey, comprising 36 communities and over 600 women, 500 of child-bearing age (15–49 years). Although the samples are small for the Secoya and Cofán, the results indicate that the Shuar and Quichua have extremely high fertility (total fertility rates[3] [TFR] of 8.8), followed by the Cofán and Huaorani at intermediate levels of 5.8 and 4.9, respectively, and the Secoya at only 3.5, or less than half the fertility of the Shuar and Quichua (Bilsborrow et al. 2007). It is interesting that a low level of fertility among the Secoya was also observed earlier by Vickers (1989), who has published widely on the Secoya. It is also useful to compare the results for the total fertility rates with those for children ever born, since the latter are based on *all* women, in contrast to age-specific fertility rates, which are derived from women of each age and hence are less reliable for specific age groups due to small sample sizes. In particular, children ever born for women aged 30–34 is a good indicator of what completed fertility is likely to be, since by that time most child-bearing has occurred. Children ever born data also differed across the five ethnic groups, from 3 to 6, varying in the same way as TFRs. The children ever born data provide results similar to those of the TFRs, further support for the TFR findings.

Explaining these differences, however, is a challenge since the Shuar have the *most* contact with market towns and commercialization of their products and the Huaorani the least, and all five ethnic groups are essentially natural fertility populations, with little knowledge of and virtually no use of modern methods of fertility regulation (Bilsborrow et al. 2007). In fact, there has been little previous research or documentation of family planning use/nonuse in indigenous populations in any part of the Amazon, an exception being the study by Hern (1991, 1992) on a small population of Shipibo in the Peruvian Amazon. Thus for our five ethnic populations, 53% of the women in unions say they want no more births, but fewer than 16% are doing anything conscious to postpone pregnancies, and most of this is use of local plants.[4] Breast-feeding is practiced commonly for 1–2 years, but varies across women, and is not intentionally used to postpone pregnancy.

---

[3]TFR is the number of live births that a woman would be expected to have during her lifetime based on current fertility levels, viz. age-specific fertility rates of women aged 15–19, 20–24, . . ., to 45–49. The TFR is the sum of the age-specific fertility rates for the seven age groups of women, each of which is based on the sum of births in a year to females of that age group divided by the number of women of that age group. It is a widely used indicator of period fertility.

[4]These numbers are actually based on preliminary tabulations from the second or survey phase of the project, based on replies from 289–298 wives of heads of households in 28 communities. The

Regarding migration, the populations are highly mobile, or at least have been up to the present. Over half of the heads of households and spouses in the eight ethnographic study communities taken together were born in other communities (of the same ethnicity) and migrated to their present residences. Quichua migrants often came from older villages to the west or south; the Shuar from the southern Amazon; the Huaorani from other villages, following traditional seminomadic proclivities; and the Cofán of Zábalo from Dureno, a Cofán community up the Aguarico River, close to Lago Agrio, the largest colonist town in the region (with a population of 34,000, according to the 2001 census of population: see Instituto Nacional de Estadística y Census 2001). In general, the massive migrations of agricultural colonists to the region from the Andes have forced the Amazonian indigenous populations to retreat eastward, further into the forest. However, the migration of indigenous families *within* the region is decreasing, as most populations are now tied to particular sites by communal legal land ownership rights and community infrastructure, such as a school, community center (*casa communal*), and occasionally a rice husker, landing strip, or pier. But there are centrifugal forces as well: as children become more formally educated and reach adulthood (and the drive of both parents and children to achieve this is strong), they may well come to be increasingly drawn to life in the regions' towns and migrate away from their communities and families more than in the past, though likely still less than the children of migrant colonists (Laurian et al. 1998).

When asked their marital or civil status, most persons over age 15 report they are married (43%), or in a *union libre* (consensual union 21%), with only 30% being single. The age at first marriage/union is generally quite young for females, being 15–18 years. Contrary to popular misconceptions in Ecuador about the indigenous populations, polygamy is almost nonexistent, with only a single case (a Huaorani man married to two sisters) among 500 women reporting.

Overall, the level of education is low as most persons who have completed formal schooling have a primary school education or less. Thus about half (48%) started but did not finish primary school, another quarter (24%) finished primary school, an eighth have no formal schooling at all, and the last eighth (11%) started but did not finish secondary school (only 3.5% completed at least secondary school). This varies only a little from community to community, though levels of education are lowest for the Huaorani. Among adults, women have less education than men and sometimes do not know Spanish. However, in the younger generation, boys and girls attend at least primary school equally. Parents of indigenous children are overwhelmingly interested in their children acquiring an education, to improve their economic prospects.

Although in the past several centuries, most of the ethnic populations in Ecuador as well as elsewhere throughout the Americas declined in population due to periods of exploitation and disease, in recent decades, most have been experiencing rapid

53% figure is still much lower than the figure for colonist women (70%) – see Bilsborrow et al. (2004).

population growth, as fertility has remained high while mortality has declined, the latter due partly to extensive immunization of children.[5] An important consequence of the demography of the indigenous populations, especially of the high fertility and population growth, is their increasing population density, which leads to rising pressures on forest resources from agriculture, hunting, and fishing. One dramatic example of this is the case of the Huaorani: over the past half century, the Huaorani population has quadrupled in size while the territory they have access to has fallen to a third of the original area. At the same time, traditional seminomadic patterns among most indigenous populations are rapidly disappearing as virtually all the land in the Ecuadorian Amazon has been titled to colonists, indigenous communities, or is in national parks or protected areas.

These two demographic tendencies – high fertility-population growth and an increase in sedentary settlement – imply increasing impacts on the natural environment, albeit concentrated in areas surrounding each community. These impacts depend on the areas cleared for agriculture and the extent of hunting (area hunted in, intensity of hunt, choice of prey, and the ecological impacts of those choices). While this discussion should not be construed as implying a Malthusian future – since population densities and the areas cleared for land use (viz. for agriculture) are still generally very low, indeed much lower than those of migrant colonists (Bilsborrow et al. 2004) – nonetheless, these demographic impacts are definitely increasing. We return to this theme later, following a review of the main economic activities of the five ethnic groups, their land use, and their hunting profiles.

8.3.2.1 A Range of Levels of Integration to the Market Economy

Before considering the use of resources (land, forests and rivers) and the implications for the environment, it is first useful to describe the range of economic patterns and associated degrees of integration to the market observed among these groups during field research. Of the five populations studied, none is a purely subsistence population – each community has people involved in the market to some degree, but this varies widely, as do the types of income-generating activities. The Shuar are oriented to wage labor and cash cropping; the Secoya to wage labor, animal husbandry involving taking out loans, and sale of timber and non-timber forest products (NTFP); the Quichua to a variety of cash crops; the Cofán to tourism and modest sales of NTFP; and the Huaorani to wage labor and sale of domestic and wild game (Table 8.2). Described below, in roughly descending order, are the groups' involvements in the market economy as of 2001.

The *Shuar*, recent migrants to the northern Amazon, are the most oriented to the market: they are engaged in the most wage labor and cash cropping, spending

[5]Data from the household surveys of rural colonist populations in 1999 and of indigenous populations in 2001 indicate a striking increase in immunization of children of both indigenous and colonist children in the Ecuadorian Amazon between the 1980s and 1990s (Pan and Erlien 2004).

Table 8.2 Summary of economic activities by indigenous group

	Shuar	Secoya	Quichua	Cofán	Huaorani
Work for petroleum companies	***	***	***		***
Cash cropping	***	***	***		
Tourism		*	*	***	*
Sale of handicrafts		*		***	***
Sale of game and fish					***
Cattle raising	***	***	*		
Timber sales	*	***	*		*

*** denotes major economic activities
* denotes minor economic activities

(based on the time allocation ethnographic study) 16.4% of their time in commercial activities and only 6.4% in subsistence activities. They raise and frequently sell domestic animals (e.g., chickens) and also sell handicrafts. They have the most frequent market transactions, purchasing food, medicine, and household items. Finally, they have the highest annual cash income (median US$241).

The *Quichua* are at an intermediate level in market integration, with the second highest participation in wage labor, cash cropping, and sale of handicrafts and domestic animals, and the highest rate of sale of forest game. However, they are next to last in cash income, and, along with the Huaorani, spend the most time in subsistence production activities (12%). This calls into question the notion that groups have to forsake subsistence production for market activities – the Quichua continue to be active in both.

The *Secoya* exhibit a mixed pattern of market integration. They spend little time in subsistence production (5.8%), have a low frequency of commercial transactions, and allocate intermediate levels of time to commercial activities (6.5%). But on the consumption side, they have many possessions at the household and community levels and relatively high use of loans, yet dietary intake patterns show only intermediate levels of market integration, and they exhibit low dependence on forest resources. However, these observed patterns may well have been distorted from normal ones by special circumstances in 2001, when the Secoya of Sewaya (the study community) were involved with Occidental Petroleum Company, which was actively exploring for oil on their lands. Occidental provided gifts to various households of manufactured goods, cash (allowing for increased consumption of purchased foods), and loans for buying cattle, creating this mixed picture of household economic portfolios observed by the study.

Finally, the *Cofán* and *Huaorani* are at the opposite end of the spectrum from the Shuar in terms of orientation to the market. Both have low frequencies of commercial transactions and use of loans, low to intermediate incomes, and spend more time in subsistence than commercial activities. The Cofán of the community of Zábalo are actively involved in ecotourism, and, although by most measures they have low market integration, they rank high in household possessions (e.g., about 4 out of 5 households own a *cocineta* (small cooking stove) and shotgun, 3 in 5 have a canoe, and half own a watch, tape player, sewing machine, chainsaw, and outboard motor). This is possible because tourism provides them with frequent connections to outsiders and access to cash and manufactured goods without having to resort to selling their labor or natural resources.

The two small Huaorani villages studied interface with the market economy mainly through the wage labor of some males for oil companies and the sale of game and handicrafts. Nevertheless, their primary orientation is still subsistence, as seen through the highest subsistence time allocation (12%), dietary reliance on forest and river resources (64% of 1,112 household-days recorded involved obtaining food from the forest or rivers), low cash income, and lack of use of external inputs in agriculture. Consumption characteristics are consistent with production findings, as the Huaorani have low levels of household possessions, low frequency of market purchases (only 1.8% of household-days), and are located far from the nearest road or market.

8.4 Cross-Cultural Hunting Patterns

As part of the ethnographic study, field researchers were trained to follow up on all hunting expeditions in their field sites by interviewing the hunters as they returned from the hunt, completing a posthunt questionnaire. As a result, over the 5 month study period, the quantity of interviews administered provides a fair indication of hunting patterns. What we find is a strong hunting orientation especially among the Cofán and Huaroani. Data on hunting frequencies indicate that the Cofán participated in the most hunts of any ethnicity (120 hunts, or a mean of 4.4 hunts per household and 0.9 hunts per capita).[6] The Huaorani had 84 hunts in the two communities combined (4.9 hunts per household, 0.8 hunts per capita). A total of 110 hunting excursions were recorded for the three Quichua villages, which amounts to 2.6 hunts per household or 0.4 hunts per capita. The Secoya community trails with 40 hunts in a community of 20 households (mean of two hunts per household, or 0.4 hunts per capita), but the Shuar community of Tiguano has the least hunting orientation of the eight ethnographic study communities, with only 10 hunts recorded (0.8 hunts per household on average, and 0.14 hunts per capita). It should be noted that perhaps the most thorough ethnographers were in Tiguano, so this low number was not due to lapses in data collection.

The ratio of the number of animal encounters divided by the number of hunting expeditions gives a rough measure of the abundance of faunal resources in various communities (Table 8.3). This ratio is lowest for the Quichua at 1.4, followed by the Secoya at 1.8 and the Shuar with 2.1. There is a large jump then to the Cofán, with a ratio of 4.1, and the Huaorani at 4.8. These last two groups apparently have a combination of a hunting orientation, hunting prowess, and a rich faunal land base.

[6]However, at the time of the study, tourism levels had dramatically declined, which is likely to have elevated rates of hunting from their usual level.

Table 8.3 Hunting frequencies and success rates, by ethnicity

Ethnicity	Number of animals encountered	Number of animals killed[a]	% of all animals killed	Number of post-hunt interviews	Ratio of encounters to hunts	Ratio of kills to hunts
Shuar	21	14	2.1	10	2.1	1.40
Quichua	158	92	13.5	110	1.4	0.84
Secoya	73	42	6.2	40	1.8	1.05
Cofán	491	357	52.6	120	4.1	2.98
Huaorani	405	174	25.6	84	4.8	2.07
Total	1,148	679	100	364	3.15	1.87

[a]This includes only reliable kill data, including sex and age of kill

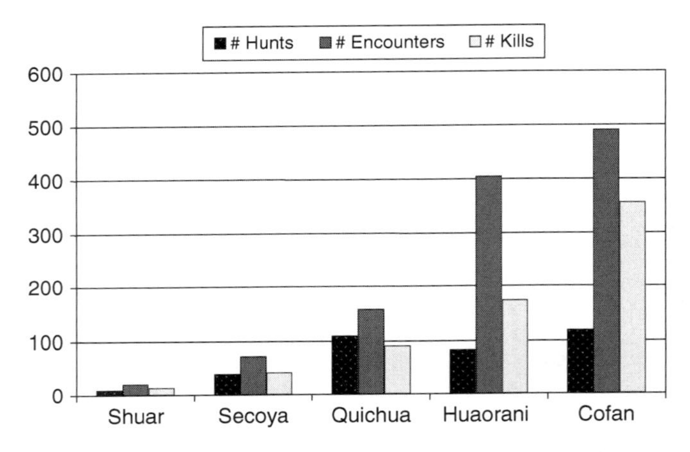

Fig. 8.3 Number of hunting excursions, animals encountered, and animals killed, by ethnicity

A glance at Fig. 8.3 illustrates this last point visually, with strikingly high levels of encounters compared to the number of hunting expeditions. Comparing the Huaorani and Cofán, it is notable that the latter are successful at killing a greater proportion of animals encountered, perhaps due to their access to the rich biodiversity of the Cuyabeno Wildlife Reserve (Table 8.3).

To compare hunting experiences of different ethnicities, it is desirable to take into account the population sizes of the communities, to compute per capita rates of actual *encounters,* in contrast to the data above on frequencies of hunts. For the Shuar ($n = 70$), the per capita rate of encounters was 0.3, while for the Quichua ($n = 267$), it was 0.6, and for the Secoya ($n = 97$), 0.75. In contrast, for both the Cofán ($n = 133$) and Huaorani ($n = 110$), it was 3.7. Thus, the ethnic groups can be grouped into three categories: the Shuar, with a very low ratio of animal encounters per capita; the Quichua and Secoya at an intermediate level, double that of the Shuar; and the Huaorani and Cofán, at a much higher level, five to six times that of the Quichua and Secoya. Note that this ranking is precisely the inverse order of the groups in terms of their levels of participation in the market economy, discussed above.

8.4.1 Prey Selection

Taking all 679 kills into account, we found a slight majority of adult animals: 52% of all kills ($n = 352$) were of fully grown individuals compared to juveniles at 47% (319). Males were more commonly killed, at 51% ($n = 346$), with females at 38% ($n = 261$) (sex data were not available for the remainder).

Besides the encounter rates of game per hunt, the composition and diversity of prey types captured is a good indicator of local forest biodiversity. Overall, rodents and lagomorphs, at 33% of total kills ($n = 227$), were the most common prey category, followed by primates (24%, $n = 161$), non-Cracid birds (15%, $n = 99$), and ungulates (9%, $n = 64$). Because primates and Cracid birds have low rates of reproduction, are large-bodied members of their taxa, and are perceived as especially desirable to eat, they tend to be the first species to be locally depleted by hunting pressure. At the other end of the spectrum, rodents and ungulates have high rates of reproduction, are less vulnerable to overexploitation, and are less desirable to eat. Therefore, we would expect to find few primates and Cracid birds around the Quichua, Secoya, and Shuar communities, and more among the Huaorani and Cofán, given their richer land bases and lower population density.

Figure 8.4 provides data on the distribution of prey types by percentage of kills and ethnicity. For the Quichua and Secoya, we found what we anticipated, with rodents (agoutis, pacas, and acouchies) being the most common types of prey captured in terms of frequency. For the Shuar, the sample size is very small, but rodents (acouchies and squirrels) also top the list, followed by ungulates (peccaries) and non-Cracid birds. For the Cofán, rodents are the most commonly killed prey

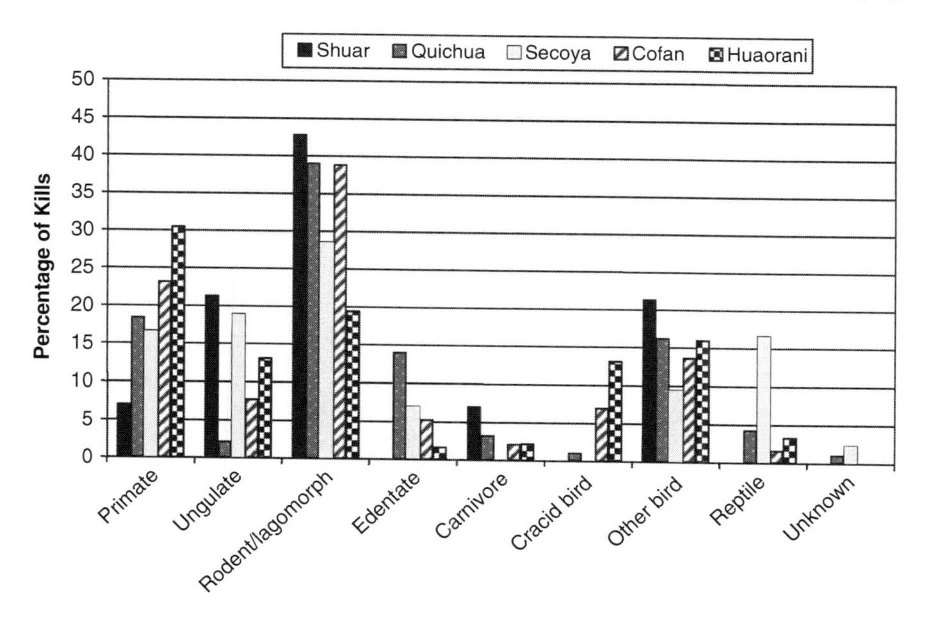

Fig. 8.4 Prey types as percentage of kills, by ethnicity

type, followed by primates. Unlike the other groups, the Huaorani kill the most primates (woolly monkey being a favorite food) in terms of the percentage of all kills, and their percentage of kills that are Cracid birds is also highest among the five ethnic groups. Indigenous groups with a relatively extensive and biodiverse territory will tend to have highly desirable prey, such as large-bodied primates and Cracid birds well represented in their hunting profiles (Redford 1992). Not only does the presence of these prey types indicate a healthy forest (e.g., given the critical ecological role these species play) but also their persistence over time in the face of hunting indicates that the populations are not being overharvested and/or there are areas that can serve as game refugia (leading to source-sink game population dynamics).

8.4.2 Technology Used

According to posthunt interviews, firearms have become the dominant method of killing game, with over 75% (512 of 679) of kills involving a gun. The next most common hunting technology is a machete, but it was used in only 8% of kills ($n = 55$). Traditional hunting tools – the blowgun and spear – were used in only 0.5% of the kills, mostly by the Huaorani. Other technologies used that were mentioned were a "stick/club," "dog," and "hands," but these were negligible.

In their classic study of hunting technologies, Yost and Kelley (1983) emphasized the wide-ranging nature of the firearm, which is able to dispatch game large and small, terrestrial, and arboreal. The distribution of prey types killed with guns is as follows: primates 28%, rodents 28%, birds 26%, and ungulates 11%. The next most common tool, the machete, was used predominantly for rodents (69%, $n = 38$), which are common in gardens and often opportunistically killed.

8.4.3 Time Allocation

We find that the Huaorani spend 3.7% of their time hunting, significantly more than the Cofán at 2.2%. The Secoya (1.2%) and Quichua (1.4%) spend, statistically speaking, the same time hunting, while the Shuar (0.7%) are the only group spending less than 1% of their time hunting. If we break down the data by gender, we see that in each of the five groups, males hunt significantly more than females, with the largest difference between sexes among the Huaorani.

8.4.4 Dietary Intake

The degree to which groups possess a hunting orientation and rely upon forest game for subsistence is demonstrated by the dietary intake data collected in the eight ethnographic study villages. Dietary checklists showed that hunted game was consumed during only 30% of the days recorded for Shuar households, while for the Quichua, this occurred on 39% of the days, followed by the Secoya on 43% of them. For the Cofán, on average, forest meat was consumed on almost half of the days (48%), and for the Huaorani, on most (70%) of the days. While some people in each of the eight communities purchase part of their food, perceptions vary widely in terms of what constitutes acceptable sustenance. In general, the Quichua informants said that if they had the money, they would buy all of their food, giving up hunting and fishing completely. Half of the Shuar said the same thing. However, the other half said they would continue to fish regardless of their ability to purchase processed foods. The Cofán and Secoya also insisted that they would still eat foods from the forest and rivers, while many Huaorani respondents said that they "always have to hunt and fish," viewing these activities as central to their identity.

8.4.5 Frequency, Distance, and Duration of Hunts

Comparing median values for hunting and fishing patterns among the eight villages, we find that median hunting frequencies (hunting trips per week) had been declining in recent years for six of the eight communities, with no change in Pachakutik and an increase in Huentaro. The latter two in fact are both splinter communities and may thus have had access to more virgin hunting lands deeper in the forest, in contrast to the other six. A strong pattern was also found among hunting frequency and the distance (mean number of kilometers hunters walked in search of game). In six of the eight villages, residents reported having to walk farther now than in the past; only in Pastaza Central was there a slight decline (the median fell from 4 to 3 km), and in Huentaro, there was no change. The decline in hunting frequencies is consistent with a decline in prey abundance, with households deciding that such an activity is not as rewarding as in the past. The increasing distances hunters have to walk for game is an even clearer indication of faunal depletion in zones close to the villages. In the past, indigenous communities in the Amazon would simply move to another part of their territory rather than contend with game depletion and longer hunting distances, but as communities have become more sedentary and territorially circumscribed, these options are much less possible.

8.5 Cross-Cultural Patterns of Land Use and Land Cover Change

Unlike the earlier sections here (except that on demography), which draw upon data only from the ethnographic study of eight villages, this section on land use utilizes data from all 36 villages covered by the household and community surveys in phase two of the project. A few results from Gray et al. (2008) are also summarized below.

The major basis of subsistence of all five indigenous groups is agriculture. All groups plant *yuca*, or manioc, and *Musa* species (mainly plantains, but also bananas and small, sweet bananas locally called *oritos*) for family consumption. For the Huaorani, there are no other primary crops, whereas all the other groups also raise additional crops, mostly to sell in the market. Thus, the Cofán grow corn, the Quichua coffee and corn, the Shuar coffee, and the Secoya coffee, cacao, and corn. The Secoya also have the highest degree of involvement in cattle production. This crop production for the market has led some Shuar, Secoya, and Quichua households to begin to also use external inputs, such as pesticides and herbicides, whereas this is not done at all by the Cofán or Huaorani.

In the survey study of 36 indigenous communities, complete data were obtained for 486 households that had one or more agricultural plots (Gray et al. 2008). The average land area in cultivation per household was 3.6 ha, which is relatively large compared to some traditional swidden systems (Beckerman 1987) but significantly less than the average amount of 15.4 ha cultivated by colonist households in the same region (Bilsborrow et al. 2004). Groups that had more involvement in the market economy were found to have significantly larger areas in cultivation: the Secoya, Shuar, and Quichua had means of 4.0–4.9 ha in cultivation per household, while the Cofán and Huaorani cultivated 2.0 and 1.4 ha per household, respectively. Interethnic differences in plot composition were also found: the Secoya cleared significantly more land than the other groups for pasture (averaging 3.4 ha per household), while the Quichua and Shuar cultivated more coffee than the other groups (1.4 and 2.1 ha/household, respectively). The area dedicated to staple crops, largely grown for subsistence, varied little across ethnicities: the Shuar, Secoya, and Quichua cleared much more land, in order to also grow cash crops and cattle, whereas all five ethnic groups had 1–1.5 ha/household in staple food crops. Thus, involvement in the market tends to *increase* the number of crop species grown, as the Shuar and Quichua were found to have the highest mean number of crop species cultivated, at 5.5 per household (Gray et al. 2008).

Not only did the amount of land cleared for cultivation differ between groups at varying levels of commercialization but so did the use of the land. The Huaorani, who, along with the Cofán, were the most subsistence-oriented, fallowed their plots for the longest period (a mean time of 2.9 years), used their current plots for significantly shorter periods (mean of 1.2 years) and maintained more distant plots (mean time of 30 min fast walk away from their dwelling), all consistent with an extensive form of agriculture. In contrast, the Secoya cleared the most forest over the previous 3 years (a mean of 4.2 ha per household) and were most likely to own cattle (70% of households vs. a mean of 14% for the other four groups of indigenous households). Because pasture exemplifies a monocrop, the Secoya were found to have the highest proportion of land monocropped, at 82% of all cleared land, with the Huaorani having the lowest proportion, at 39%.

The Huaorani and Cofán, who use the smallest cultivated areas of the groups studied and focus on subsistence crops, are correspondingly the most reliant on *nonagricultural* activities, some of which have land use implications. The majority of the Huaorani households studied (85%) had some member of the household (always male) employed for wages at some time during the 12 months before the survey interview, most engaged in manual labor for an oil company (e.g., clearing forest, assisting in seismic exploration, or cleaning up oil spills). In addition, 50% of the Huaorani households surveyed had someone who had hunted in the previous week, and 58% had sold handicrafts at some time during the previous 12 months, though earning little income. Cofán households also sold handicrafts and products made from foraged forest resources, such as medicinal plants.

Patterns of land use indicate that those groups who participate in commercial agriculture tend to clear larger areas (because all groups, as noted, maintain similar areas for subsistence food crops), fallow for shorter periods, and cultivate more crop varieties (cash crops as well as the same staple crops raised by others for subsistence). An exception is cattle raising, as seen among the Secoya, which involves clearing large areas and greatly simplifying the plant community as cattle grazing reduces the landscape to a few species of grasses able to withstand intensive culling.

8.6 Discussion

In this paper, we provide a cross-cultural example of the importance of both the sociocultural and the economic contexts in understanding indigenous patterns of forest use in the Ecuadorian Amazon. Among the sample of eight communities spanning five ethnic groups, we found important similarities as well as differences. In all cases, the five indigenous populations are young, with half or more under age 20, are rapidly growing due to high fertility and declining mortality, and are increasingly tied to nucleated settlements providing schooling and other amenities. They are all also engaged in mixed economies involving both subsistence and market production, and all hunt and fish as well, albeit to greatly varying degrees. However, there are wide differences in the extent of dependency of each ethnicity on these various economic activities, with direct implications for the impacts they have on the biologically rich environment they live in. In the paragraphs that follow, we consider the differences in economic activities (agriculture/land use, wage labor, and hunting) and their environmental implications among the five ethnicities, including noting possible trade-offs.

First, as we have seen, there is a range of hunting patterns among the five indigenous groups studied, with the Shuar, recent migrants to the northern Ecuadorian Amazon and arguably the group most integrated to the market during the data collection period, having the least emphasis on hunting – the fewest excursions, the lowest rate of forest game consumption, the least time spent hunting, and the greatest proportion of prey consisting of common animals such as rodents. The Shuar community (Tiguano) in the ethnographic sample is located near a major oil road, and an oil company was actively looking for oil during the ethnographic study, providing wage labor opportunities and gifts to the community. As a result, the Shuar became even more involved with external market forces at the time, leading some families to permanently migrate to a nearby town shortly after the study period. Those remaining continue to focus on commercializing crops and thereby have a strong impact on the environment in terms of cleared area.

Hunting data on prey encountered and killed, mediated by cultural dietary preferences, reveal connections between land accessible to the community and faunal diversity. Among indigenous communities with greater access to land than the Shuar, especially the Huaorani, primates and other large-bodied animals with low rates of reproduction are exploited at higher rates than they are in communities with less land. This conforms to our expectation that these prey categories, indicative of higher faunal diversity, are found only within hunting distance of communities with a large land base and therefore lower risk of overexploitation. The Huaorani thus have an impact through hunting since their source of protein is mostly hunted animals and fish, and they focus on certain prized game species (e.g., large primates and Cracid birds). At the same time, they purchase and sell little in the market, clear only small areas for subsistence agriculture, and through their land use and fallow practices have much smaller impacts on the land than the other groups, dispersed over a larger area.

At an intermediate level of market integration in 2001 were the Secoya and Quichua, although they are highly heterogeneous. The Quichua communities, which combine market and subsistence pursuits, have a long history of contact with outsiders and, accordingly, have developed great ability to adapt to changing circumstances. As the most numerous of the indigenous groups in Ecuador's Amazon, and with very high fertility, their population pressure on land and resources is rising. While they still maintain a cultural emphasis on hunting for subsistence, as seen by the number of hunts recorded over the study period, they also have the lowest encounter rate of game of the five groups, in part due to the already relatively low biodiversity, and hence rarely capture animals other than the more common rodents and lagomorphs. The Secoya have been most involved with oil companies for several years, receiving money and loans at the household and community levels in return for the right to prospect for oil, and they were also encouraged to raise cattle. This led to a jump in land clearing easily discernible in satellite images. The Secoya are aware of their shrinking land and resource base. During the 2001 ethnographic study, Secoya residents undertook a number of hunts – almost as many as the Huaorani and Quichua – but, like the Quichua, experienced one of the lower rates of animal encounters, killing mostly rodents.

Lastly, the Cofán village studied was a bit of a *sui generis* case during the study period, though in general, the Cofán tend to be similar to the Huaorani in their dependence on the forest more than the market. Located in the Cuyabeno Wildlife Reserve in a remote eastern Amazonian area near the border with Peru, the community of Zábalo was founded by a group of Cofán who moved downriver to escape intrusions of petroleum companies and migrant colonists that were occurring around the Cofán community of Dureno near Lago Agrio. They are highly dependent on tourism revenues, though they supplement this source of income with hunting, fishing, and small agricultural plots or vegetable gardens. The abundance of wildlife around Zábalo makes their hunting forays successful, with the Cofán reporting the highest number of kills during the 2001 study period. However, with

Plan Colombia scaring away tourists, the Cofán have had to resort to clearing more forest to plant additional gardens during the 2001 study period and subsequently.

It is noteworthy how the different ethnic groups can be viewed as trading off different forms of economic activity with different implications for resource use. Thus, the Shuar focus on market-oriented activities, engage in little hunting, and clear large areas of forest for cattle and crops. They thus have intense impacts on the environment, similar to those of colonists, albeit on a smaller scale. At the other end, the Huaorani and Cofän clear only very small areas and practice long fallow and polycropped agriculture, which is better for forest recovery. Nevertheless, their dependence on hunting and their hunting prowess ensures that they do have impacts on various animal species (and indirectly on the animal and plant species linked to these prey species) over fairly large areas. In between are the Quichua and Secoya, the former highly heterogeneous in economic activities from one village to another, and the latter recently evolving towards the Shuar in adopting cattle raising.

As for which populations have the greater impacts, there is no definitive answer, although it appears that it was the Shuar and Secoya in the 2001 study period. However, impacts depend on hunting and agricultural practices, which are linked to their evolving consumption aspirations and therefore needs for income. The key issues then are how fast are their consumption desires changing toward market goods, how fast are their diet preferences changing, and what can be anticipated in the future. But *in all cases, for all ethnicities*, their rapidly growing populations and sedentarization ensure that the *environmental impacts will continue to grow*.

In sum, the five study populations are all involved in the market economy to varying degrees, though substantial diversity exists both inter and intraculturally in land use and conservation, from patterns of agricultural clearing, cropping, and fallowing, to reliance on forest and river fauna. Hence, policies to promote conservation or "sustainable development" among the Shuar should be quite different from those for the Huaorani. For example, while the Shuar have more land cleared in homogenous plots of cash crops, the Huaorani have smaller, polycropped plots of crops, even using fruit trees in their *chacras* for food for years after ceasing cultivation. The oversimplified dichotomy of the "ecologically noble savage" versus *Homo devastans* is thus incorrect on many levels; rather the picture that emerges is a complex diagram of different pressures, opportunities, cultural preferences and values, and historical constraints, which, taken together, influence the choices people make about how to use land and resources. As these factors change, so will the choices people make. Conservation and development policies thus need to be people-, place-, and period-specific, to embrace the human dynamism and human and ecological diversity.

8.7 Conclusions

The Amazon rainforests of Ecuador, covering only 138,000 km^2, house an extraordinary richness of biological *as well as* cultural diversity. Recent estimates of the number of indigenous people living in the region range from Irvine's (2000) estimate of 104,000 to a 2001 census estimate of 162,868 (Instituto Nacional de Estadística y Censos, 2001). Although this may appear to be a small number in absolute terms, this population is nearly as numerous as the indigenous population of the *entire* Brazilian Amazon, which covers 45 times the land area. Furthermore, 75% of the Ecuadorian Amazon is claimed by the native peoples who have long lived there (Irvine 2000: 22) and perhaps half of this has by now been legally titled to them (though subsurface rights, as throughout Latin America, pertain to the State, creating persistent policy conflicts whenever petroleum is discovered in new places in the region). Conservation efforts require interdisciplinary approaches that integrate natural and social sciences. As we have seen in this chapter, insights and tools from demography, anthropology, economics, and geography all play important roles in understanding the underlying processes behind deforestation and changing patterns of land and resource use in the study region, in which indigenous populations are involved, together with migrant colonists, petroleum companies, and the State through its development policies and creation of protected areas and national parks.

Demographic aspects, including high fertility and population growth and population redistribution through migration (but declining at a community level over time due to increasingly sedentary settlement), directly raise person/land ratios, which increase human impacts on the environment. Recent research on human–environment interactions in anthropology and geography, however, demonstrates that, while population growth is important, the *context* in which the population growth occurs is critical, necessitating a more comprehensive approach which recognizes that human impacts are mediated by sociocultural, technological, political, economic, geographical, and biophysical characteristics (Blaikie and Brookfield 1987; Bilsborrow 1987; Geist and Lambin 2002; Entwisle and Stern 2005).

Throughout Amazonia and not just Ecuador, indigenous peoples find themselves and their lands increasingly circumscribed by new land uses and land users. Commercial enterprises extracting raw materials, energy, and agro-business are gaining access to more and larger tracts of tropical forests, often bolstered by state policies aimed at stimulating development in the region. More and more rainforest is thus being expropriated for colonization, cash cropping, cattle production, and even urban areas (Browder and Godfrey 1997; Bilsborrow and Vallejo 2002). Far from being the timeless, isolated, and static romanticized natives, indigenous peoples in Ecuador and elsewhere are engaging and grappling with forces of change, making tradeoffs, seeking opportunities, and adjusting to outsiders and outside economic forces, with varying but generally increasingly significant environmental impacts as their populations grow and contacts with market forces expand.

Studies that examine these changing patterns of human/environment interactions in more detail are needed in Ecuador and elsewhere to (a) increase our understanding of how cultural values and perceptions of the natural environment translate into land and resource use practices; (b) document further the ecological impacts of economic and political changes (viz. increased market participation, the switch in Ecuador from its national currency, the *sucre*, to the dollar at the end of 1999, and changes in land tenure); (c) test theories of the impacts of population dynamics on forest cover and quality, including neo-Malthusian theories of agricultural extensification (Malthus 1798) and Boserupian theories of increasing land use intensification (Boserup 1965); (d) inform nuanced conservation policy that is context-specific and culturally appropriate; and (e) foster more realistic perceptions of indigenous people in contrast to romanticized notions, emphasizing our common humanity rather than portraying them as exotic curiosities.

As we have seen, key factors in understanding variations in land use patterns across cultural groups include differences in demographic characteristics and dynamics; proximity to infrastructure, especially roads and towns; and the influence of markets, not only economically but also culturally; the prevalence and types of available wage work; and the strength of cultural ties to the land. While the importance of each of these sets of factors varies with time and context, and the interplay of cultural, economic, demographic, and biophysical factors is complex, efforts to tease out the relationships both quantitatively and in more contexts/regions are needed to point towards better policies for dealing with the ongoing disappearance of biological and cultural diversity in the Amazon.

Acknowledgments We are grateful for funding from the National Institutes of Health (RO1-HD38777); for comments on a previous draft from the editors and anonymous reviewers; for the collaboration of our two subcontractors, Fundación Ecociencia and Centro de Estudios sobre Población y Desarrollo Social (both in Quito, Ecuador); and to the Spatial Analysis Unit of the Carolina Population Center, particularly Brian Frizzelle. Finally, we thank our principal Ecuadorian colleagues, Ana Oña and Francis Baquero, for assistance in fieldwork; our colleague, Bruce Winterhalder for suggestions in the design phase of the ethnographic project; Jason Bremner and Clark Gray for data cleaning and analysis; all the ethnographers who carried out the intensive fieldwork; and the residents of the 36 indigenous study villages for their collaboration.

References

Albuja L, Cáceres F, Almendáriz A (2000) Diagnóstico ambiental en el sector ampliado del convenio entre el Ministerio del Ambiente y la Comunidad Cofán de Zábalo, Reserva de Producción Faunística Cuyabeno. Informe Final Convenioo, Proyecto PETRAMAZ-Escuela Politécnica Nacional, Quito, Ecuador

Beckerman S (1987) Swidden in Amazonia and the Amazon Rim. In: Turner BL II, Brush SB (eds) Comparative farming systems. Gilford, New York, pp 55–94

Bilsborrow RE (1987) Population pressures and agricultural development in developing countries: a conceptual framework and recent evidence. World Dev 15(2):183–203

Bilsborrow RE, Vallejo L (2002) Urbanization, infrastructure change and development in the Ecuadorian Amazon: from the ground up. Presented at Population Association of America annual Meeting, Atlanta, Georgia, 8–11 May 2002

Bilsborrow RE, Barbieri AF, Pan WK (2004) Changes in population and land use over time in the Ecuadorian Amazon. Acta Amazon 34(4):635–647

Bilsborrow RE, Bremner J, Holt FL (2007) El Comportamiento Reproductivo de Poblaciones Indígenas: Un Estudio a la Amazonía Ecuatoriana. Centro de Estudios de Población y Desarrollo Social and United Nations Fund for Population Activities, Quito, Ecuador

Blaikie P, Brookfield H (1987) Land degradation and society. Metheun, New York

Borgerhoff-Mulder M, Caro TM (1985) The use of quantitative observational techniques in anthropology. Curr Anthropol 26:323–335

Boserup E (1965) The conditions of agricultural growth: the economics of Agrarian change under population pressure. Aldine, New York

Brosius JP (1997) Endangered forest, endangered people: environmentalist representations of indigenous knowledge. Hum Ecol 25(1):47–69

Browder J, Godfrey J (1997) Rainforest cities. Columbia University Press, New York

Cabodevilla MA (1989) Memorias de Frontera: Misioneros en el Río Aguarico (1954–1984). CICAME, Ecuador

Cabodevilla MA (1997) La Selva de Los Fantasmas Errantes. CICAME, Ecuador

Califano M, Gonzalo JA (1995) Los A'i (Cofán) del Río Aguarico, Mito y Cosmovisión. Serie Pueblos del Ecuador, Ediciones Abya Yala, Quito, Ecuador

Cerón MC (1995) Etnobiología de los Cofanes de Dureno. Serie Pueblos del Ecuador, Ediciones Abya Yala, Quito, Ecuador

Conklin BA, Graham LR (1995) The shifting middle ground: Amazonian Indians and eco-politics. Am Anthropol 97:695–710

De La Torre S, De La Torre L, Oña AI (2000) Diagnostico Ambiental, Cultural y Socio Económico de las Cabeceras del Río Aguas Negras, en la Reserva de Producción Faunística Cuyabeno y del Centro Siecoya Remolino. PETRAMAZ (Proyecto de Gestión Ambiental: Explotación Petrolífera y Desarrollo Sostenible en la Amazonía Ecuatoriana), European Commission and Ministry of Environment of Ecuador, Quito, Ecuador

Entwisle B, Stern PC (2005) Population, land use and environment: research directions. National Research Council, Washington DC

Fisher WH (2000) Rainforest exchanges: industry and community on an Amazonian frontier. Smithsonian Institution Press, Washington DC

Food and Agriculture Organization of the United Nations (2005) State of the World's Forests 2005. UN Food and Agriculture Organization, Rome

Geist HJ, Lambin EF (2002) Proximate causes and underlying driving forces of tropical deforestation. BioScience 52:143–150

Godoy RA (2001) Indians, markets and rainforests: theory, methods, analysis. Columbia University Press, New York

Gray CL, Bilsborrow RE, Bremner JL, Lu F (2008) Indigenous land use in the Northern Ecuadorian Amazon: a cross-cultural and multilevel analysis. Hum Ecol 36:97–109

Hendricks JW (1988) Power and knowledge: discourse and ideological transformation among the Shuar. Am Ethnol 15:216–238

Hern WM (1991) Effects of cultural change on health and fertility in Amazonian Indian societies: recent research and projections. Popul Environ 13(1):23–43

Hern WM (1992) Polygyny and fertility among the Shipibo of the Peruvian Amazon. Popul Stud 46:53–64

Holt FL, Bilsborrow RE, Oña AI (2004) Demography, household economics, and land and resource use of five indigenous populations in the Northern Ecuadorian Amazon: a summary of ethnographic research. Carolina Population Center Occasional Paper, University of North Carolina, Chapel Hill, NC

Instituto Nacional de Estadística y Census (2001) Censo de Población y de Vivienda – Resultados. Electronic document. http://www.inec.gov.ec

Irvine D (2000) Indigenous federations and the market: the Runa of Napo, Ecuador. In: Weber R, Butler J, Larson P (eds) Indigenous peoples and conservation organizations: experiences in collaboration. World Wildlife Fund, Washington DC, pp 21–46

Kimerling J (1991) Amazon crude. Natural Resources Defense Council, Washington DC

Laurian L, Bilsborrow R, Murphy L (1998) Migration decisions among settler families in the Ecuadorian Amazon: the second generation. Res Rural Sociol Dev 7:169–195

Lu FE (1999) Changes in subsistence patterns and resource use of the Huaorani Indians in the Ecuadorian Amazon. Doctoral Dissertation, University of North Carolina at Chapel Hill

Malthus TR (1798) An eassy on the principle of population, London

Myers N (1988) Threatened biotas: hotspots in tropical forests. Environmentalist 8:1–20

Pan W, Bilsborrow RE (2005) The use of a multilevel statistical model to analyze factors influencing land use: a study of the Ecuadorian Amazon. Global Planet Change 47:232–252

Pan WK, Erlien C (2004) Impact of population-environment dynamics on human health in the Ecuadorian Amazon. Presented at Annual Meeting of American Public Health Association, Washington DC

Pichón FJ (1997) Colonist land-allocation decisions, land use, and deforestation in the Ecuadorian Amazon Frontier. Econ Dev Cult Change 45(4):707–744

Pitman N, Terborgh J, Silman M, Nunez P, Neill D, Ceron C, Palacios W, Aulestia M (2002) A comparison of tree species diversity in two upper Amazonian forests. Ecology 83:3210–3224

Redford KH (1990) The ecologically noble savage. Orion (Summer):25–29

Redford KH (1992) The empty forest. BioScience 42:412–422

Romoleroux K, Foster R, Valencia R, Condit R, Balslev H, Losos E (1997) Especies Leñosas Encontradas en Dos Hectáreas de un Bosque de la Amazonía Ecuatoriana. In: Valencia R, Balslev H (eds) Estudios Sobre Diversidad y Ecología de Plantas. Pontífica Universidad Católica del Ecuador, Quito, Ecuador, pp 189–215

Rudel TK (1983) Roads, speculators, and colonization in the Ecuadorian Amazon. Hum Ecol 11:385–403

Rudel TK, Bates D, Machinguiashi R (2002) Ecologically noble amerindians? cattle ranching and cash cropping among the Shuar and Colonists in Ecuador. Lat Am Res Rev 37:144–159

Terborgh J (1999) Requiem for nature. Island, Washington DC

Valencia R, Balslev H, Paz y Miño G (1994) High tree alpha-diversity in Amazonian Ecuador. Biodivers Conserv 3:21–28

Vickers W (1989) Los Sionas y Secoyas. Su Adaptación al Ambiente Amazónico, Abya-Yala, Quito, Ecuador

Yost JA, Kelley PM (1983) Shotguns, blowguns, and spears: the analysis of technological efficiency. In: Hames RB, Vickers WT (eds) Adaptive responses of native Amazonians. Academic, New York, pp 189–224

Chapter 9
Human Demography and Conservation in the Apache Highlands Ecoregion, US–Mexico Borderlands

L.J. Gorenflo

9.1 Introduction

During the second half of the twentieth century, the area along the United States (US)–Mexico border evolved from a remote frontier to a region experiencing considerable development. As a consequence of this development, between 1970 and 2000 parts of this *Borderlands* region witnessed some of the highest rates of population growth documented in either of the two host countries (Canales 1999; Lorey 1990, 1993), with rapid growth projected to continue in much of the region through at least 2020 (Peach and Williams 2000). Among the impacts of development in the Borderlands has been the destruction of broad tracts of natural habitat and the biodiversity that relies upon it. As conservationists and researchers come to recognize more fully the rich variety of life present in this region, they have increased efforts to conserve key components of its biodiversity for future generations.

In an attempt to define meaningful geographic units for conserving biodiversity in the Borderlands and neighboring areas, biologists at The Nature Conservancy (a US-based nongovernmental organization whose central mission is biodiversity conservation) have divided the region into six different ecoregions that straddle the border, along with two that lie immediately south of the border. Comprising large land areas designated by climate, vegetation, geology, and other ecological and environmental variables (The Nature Conservancy 2000), these ecoregions represent the main geographic units of conservation planning for the Conservancy. Resulting ecoregional plans identify conservation targets, prioritize these targets, and define portfolios of sites whose conservation will ensure the persistence of essential biodiversity in each ecoregion. Although such plans invariably acknowledge the important role of human impacts in reducing biodiversity, they tend to focus limited explicit attention on humans and their activities. This chapter examines human demographics and development in one Borderlands ecoregion – the Apache Highlands – in an attempt to improve our understanding of past, present, and future human impacts to conservation in an area recently the focus of ecoregional planning by The Nature Conservancy.

With a general emphasis on human population in the Apache Highlands Ecoregion, the ultimate aims of this study are three: (1) to document geographic

R.P. Cincotta and L.J. Gorenflo (eds.), *Human Population: Its Influences on Biological Diversity*, Ecological Studies 214,
DOI 10.1007/978-3-642-16707-2_9,

patterns of population in the ecoregion; (2) to explain recent demographic change in the ecoregion; and (3) to identify those aspects of Apache Highlands demography that provide insights on biodiversity conservation. The study begins with a brief historic overview of human occupation in the Apache Highlands, to provide a foundation from which to understand its demographic evolution as well as the current geographic arrangement of population and development in this ecoregion. It then examines recent population levels and settlement patterns in the Apache Highlands, describing the geographic distribution of human population and exploring possible underlying causes leading to the current distribution. The paper examines proposed conservation sites in terms of population density and rates of change in an attempt to identify the utility of population density and average annual rate of change as indicators of biodiversity disruption. It closes by considering the interrelationship between biodiversity conservation and human demographics in this portion of the US–Mexico Borderlands, proposing both a population density threshold beyond which high biodiversity was unlikely and identifying localities where biodiversity might be compromised in the future.

9.2 An Overview of Human Occupation in the Apache Highlands

The Apache Highlands Ecoregion covers roughly 12 million ha in southeastern Arizona and southwestern New Mexico in the United States and northeastern Sonora and northwestern Chihuahua in Mexico (Fig. 9.1). Lying at the northern end of the Sierra Madre Occidental, the Apache Highlands is distinguished from this extensive mountain range (and ecoregion of the same name) to the south by the presence of small, isolated mountain ranges called *sky islands* (Marshall et al. 2004). The physical geographic and ecological diversity of this ecoregion includes elevations in excess of 2,100 m separated by expanses of relatively flat desert grasslands and riparian corridors. Biodiversity in the Apache Highlands Ecoregion includes roughly 110 mammal species, 265 bird species, 75 reptile species, 190 snail species, and 2,000 plant species and features both endemic and endangered plants and animals. The recent ecoregional planning effort comprises one of a series of activities to help conserve key portions of the biodiversity in the Apache Highlands.

Humans have played a role in Apache Highlands ecology for millennia. Archeologists have found evidence of human habitation in this ecoregion dating as early as 15,000 before the present (BP), associated with a time period that prehistorians call *Paleo-Indian* (Cordell 1997). Although the defining characteristic of Paleo-Indian is large game hunting, the small groups of hunter-gatherers from this period of prehistory probably used a range of other foods as well. Beginning around 9500 BP and continuing until about 1300 BP (AD 700), inhabitants of the Apache Highlands and the rest of the greater southwest followed an adaptive strategy called *Archaic* – characterized by mobile bands of hunter-gatherers that exploited a particularly broad range of plant and animal resources that occurred

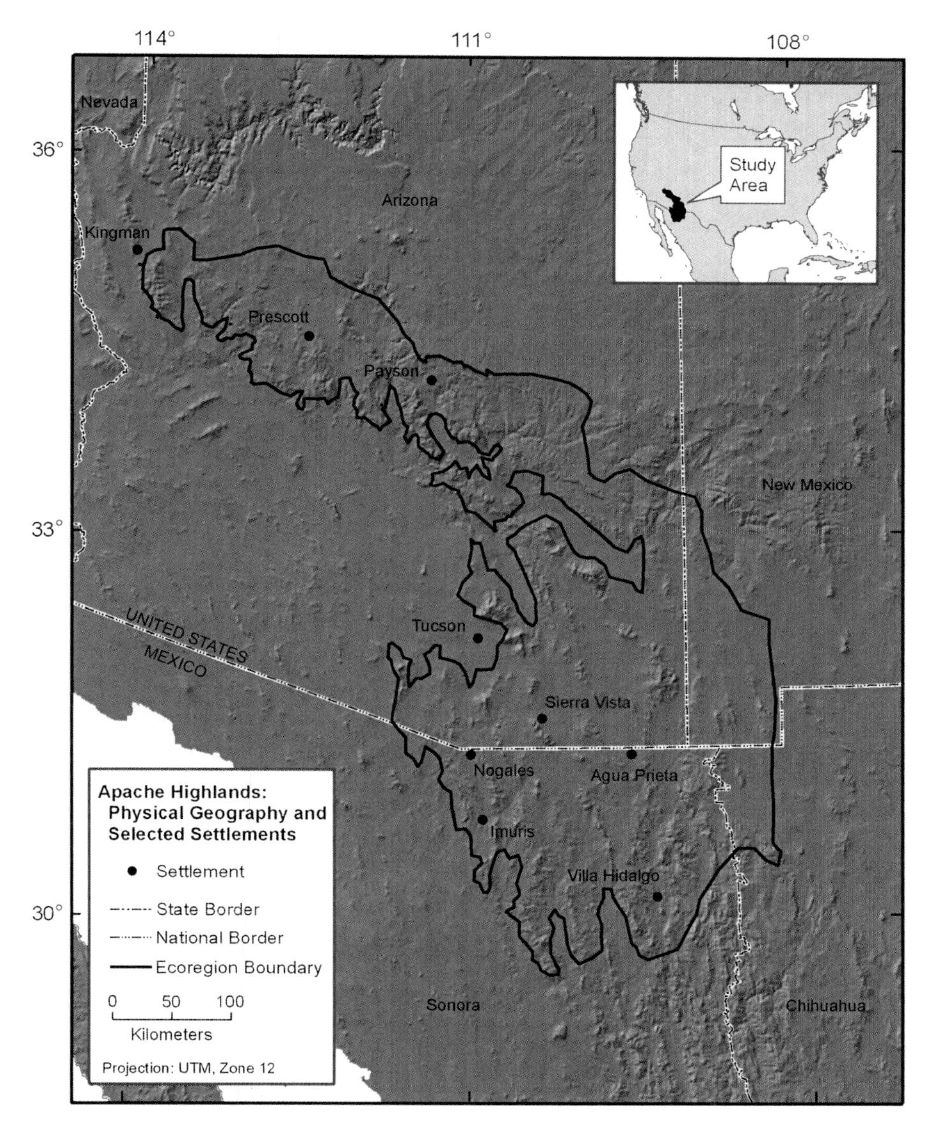

Fig. 9.1 The physical geography and main communities in the Apache Highlands Ecoregion (data sources: Environmental Systems Research Institute 1992; National Aeronautics and Space Administration 2004)

naturally in the areas where they lived. The intimate familiarity with plants included use of the early ancestors of corn, whose remains appeared with increasing frequency in archeological sites in the region beginning about 4000 BP (Martin 1979). After 1300 BP (AD 700), reliance on agriculture increased and continued beyond the arrival of Europeans in the sixteenth century. Archeologists refer to the agricultural adaptation to mountainous regions in southeastern Arizona, southwestern New Mexico, northeastern Sonora, and northwestern Chihuahua as *Mogollon*, one of the most important and well-studied prehistoric southwestern cultural traditions (Haury 1936, 1985). With its roots stretching back to 4000 BP, Mogollon continued until about 600 BP (circa AD 1400) and was characterized by small sedentary villages, initially built in defensive localities and in later times in more accessible and agriculturally productive localities. Although the prehistory of much of the Apache Highlands Ecoregion is relatively well understood, research does not include intensive archeological surveys over large areas that would provide a basis for reliable population estimates. Given the nature of subsistence and settlement, and what is known of early historic demography in the area, population would have been sparsely distributed and probably totaled a few tens of thousands.

The dominant strategies of cultural adaptation found prehistorically in the Apache Highlands continued into the historic indigenous socio-cultural systems of the area. The most recent overview of tribal peoples in this part of North America lists several groups with historic ties to the ecoregion (Ortiz 1983): Chiricahua Apache, Eudeve, Jano, Jocome, Lower Pima, Ópata, Tohono O'odham, Upper Pima, Yavapai, Walapai, and Western Apache. The majority of these peoples lived in small, seminomadic groups and survived through hunting and gathering, often supplemented by small-scale agriculture (Basso 1983; Dunnigan 1983; Ezell 1983; Gifford 1936; Goodwin 1942; Griffen 1969; Hackenberg 1983; Hinton 1983; Opler 1941; Sauer 1934). The Tohono O'odham and (especially) Pima were exceptions to this, living in larger sedentary settlements and relying more heavily on irrigated agriculture. However, the core geographic areas for these peoples tended to lie to the west in the Sonoran Desert, the Pima in particular living near larger rivers and broad alluvial plains more conducive to large-scale agricultural activity. The Ópata, occupying what is today the south-central portion of the Apache Highlands, also relied heavily on irrigated agriculture, though settlement tended to consist of small hamlets (Hinton 1983). Lack of data, coupled with considerable demographic and cultural disruption that accompanied the arrival of Euro-Americans, limits our understanding of historic aboriginal population in the ecoregion. Scholars generally agree that the total inhabitants in Pima Bajo (roughly corresponding to the current state of Sonora) and Pima Alto (northernmost Sonora and the southern half of Arizona) probably totaled fewer than 40,000 at the time of European contact (Pennington 1980; Sauer 1935). Adding a fraction of the Ópata population for that portion of their homeland included in the Apache Highlands (see Gerhard 1993) would increase the population estimate for what is now northern Sonora and southern Arizona to perhaps 50,000, with the population of the ecoregion portion of this area substantially less.

The first Europeans in the Apache Highlands Ecoregion were Spanish explorers who entered the region in the 1530s and early 1540s from both the north and south, primarily in search of personal wealth (Hartmann 1989; Spicer 1962). Initial Spanish colonization of the southern and eastern parts of the state of Sonora occurred by the early seventeenth century and expanded into all but the western part of that state and southern Arizona by 1710 (Gerhard 1993; Jackson 1998;

Ortega 1993; Spicer 1962). The first Spanish settlements in the region consisted primarily of missions, small towns, and *presidios* (military garrisons), with early colonists in the Apache Highlands area engaged primarily in mineral (silver) mining, along with small-scale ranching and agriculture (West 1993). Most settlements were small and widely scattered, as conflicts with indigenous peoples, the harsh natural environment, and limited economic opportunities hampered more substantial colonization. Although Hispanic expansion northward continued following the Mexican War of Independence in 1821, the nonindigenous population remained fairly small, totaling perhaps 15,000 by 1821 for the state of Sonora as a whole (Gerhard 1993). Disease and conflict decimated the indigenous population of the Apache Highlands and other parts of this general region during the first centuries of European occupation, though the greatest impacts initially occurred in the southern part of the ecoregion. The Native population of Sonora alone declined to fewer than 8,000 in the early nineteenth century (Gerhard 1993), producing a substantial net decline in total regional demographics.

Beginning in the mid-1850s, Anglo-Americans from the United States expanded into the territories of Arizona, California, and New Mexico, newly acquired in the Mexican–American War. Settlement occurred slowly along a military front, in the wake of indigenous hostilities, based on an economy of mining, ranching, and agriculture similar to the pattern seen in Sonora the preceding two centuries (Sheridan 1995). About the same time, settlement in Sonora began to expand into western parts of the state – in part drawn by discoveries of gold in those areas and driven by the exhaustion of mines in the eastern mountains, and in part made possible through greater control over Native hostilities (West 1993). Colonization increased after the 1880s, particularly in Arizona and New Mexico following the cessation of most indigenous hostilities, although communities such as Tucson received many more in-migrants than did the more mountainous Apache Highlands a short distance to its east. Water control began on a larger scale in the late nineteenth century, expanding with the construction of large dams in the United States and Mexico during the 1920s and 1930s and providing a more reliable foundation for agriculture and settlement. Finally, the establishment of rail links with other parts of Mexico and the United States promoted economic growth in the southwestern United States and northwestern Mexico (Sanderson 1981) and the fortunes of communities on the rail lines, though the focus of development once again was on less mountainous portions in both countries beyond the borders of the Apache Highlands Ecoregion. Despite a resurgence of mining in eastern Sonora near the end of the nineteenth century (West 1993), most development and new settlement occurred outside the bounds of the ecoregion on both sides of the border.

Well into the twentieth century, agriculture, ranching, and mining continued to support much of the human habitation in the Apache Highlands Ecoregion. Nevertheless, population remained relatively small as development on both sides of the border focused on other areas. Population and economic growth surged generally on the Mexican side of the border early in World War II, as exports to the United States grew markedly (Sanderson 1981). Similarly, population growth occurred in Arizona following World War II and the accompanying expansion of manufacturing (primarily electronics) to supplement other forms of economic activity. Settlement in the small sections of Chihuahua and New Mexico that lie within the ecoregion continued to experience relatively slow growth and development, remaining generally rural and sparsely populated. In contrast, population in the Arizona and Sonora portions of the Apache Highlands grew more rapidly. In the case of the former, population in the United States continued to relocate to the Sun Belt states, both for employment and retirement. In the case of the Sonora portion of the Apache Highlands, recent population growth has followed the emergence of economic opportunities on the American side of the border and the meteoric growth of manufacturing and assembly industry (the *maquiladoras*) on the Mexican side of the border (see Alegría 1992; Canales 1999; Sable 1989). Migrants from elsewhere in Mexico relocated to exploit these economic opportunities, often fleeing undesirable social, economic, and environmental conditions in their former homes (National Heritage Institute 1998).

9.3 Population in the Apache Highlands Ecoregion at the End of the Twentieth Century

Although the discussion in the preceding section necessarily relies largely on information from more general geographic areas rather than solely the Apache Highlands, it provides a basic understanding of human demography in this region. Due to a combination of factors, the indigenous population of the ecoregion was never very dense – owing mainly to sparse settlement by hunter-gathers and agriculturalists living outside of more productive broad alluvial plains, such as those to the west in the Sonoran Desert (Gorenflo 2002). Euro-American colonization was slow in the face of conflicts with natives and challenges of the harsh natural environment, though demographically noteworthy for the tragic decimation of indigenous populations (largely from introduced diseases) which reduced them to small fractions of their original sizes. Subsequent settlement in northwestern Mexico and the southwestern US increased, substantially during the late twentieth century, but focused on other areas in the general region outside the more mountainous Apache Highlands. Nevertheless, population in the ecoregion grew, primarily after 1950 and particularly near the end of the twentieth century. By 1990 ecoregion population approached 569,000. By 2000, population of the region exceeded 797,000 – an increase of 40% in only 10 years and totaling more than the entire population of Arizona only five decades earlier (US Bureau of the Census 1996).[1]

[1]Estimates of ecoregion population used geographic information system technology to calculate the proportion of block groups and áreas estadísticas básicas (discussed below) lying within the ecoregion boundary and applied this percentage to the population of each in 1990 and 2000. For geographic units only partially within the boundary, this method assumes uniform population density to estimate the number living inside the ecoregion. Although such an assumption is incorrect, because the geographic units employed are small the errors introduced to the estimate similarly will be small.

One can examine recent population data for the Apache Highland Ecoregion in geographic units of several different sizes. Counties in the United States and their geographic counterparts in Mexico, *municipios*, provide a good general sense of the distribution of ecoregion population and how this population has changed over time (Table 9.1). Although geographic units this large are of limited use when examining most conservation sites, due to the lack of precise placement of population with respect to site locations, the general availability of demographic statistics for larger units enables one to identify broad demographic patterns that provide important insights on the interface between human demographics and localities of importance for conservation. The population in counties and municipios partially or totally within the ecoregion (or outside the ecoregion but near the boundary) ranged widely in size from more than 3 million to fewer than 1,000 in 2000, the year of the most recent available decennial census in both Mexico and the United States. At the county/municipio level, one sees a distinction between demographics in the United States and Mexico portions of the ecoregion; much larger recent populations occur in the former, though the population of Nogales Municipio immediately south of the border was nearly 160,000 in 2000. Recent population change in the Apache Highlands Ecoregion similarly has varied considerably across its geographic extent. Once again, a distinction is evident between those portions of the ecoregion north and south of the border. Between 1990 and 2000, population grew in every county in the US portion of the ecoregion except one. In contrast, more than half of the 34 Mexican municipios at least partially within the ecoregion lost population during the same time period. Eighteen of the thirty-two municipios in the ecoregion that existed in 1950 (two were created after that date, split off from other municipios) actually lost population over the second half of the twentieth century. This reduction in population likely is a consequence of declining economy in the mountains of Sonora and Chihuahua, running counter to demographic trends throughout much of the Borderlands during the same period. Only one of the 15 US counties in the ecoregion lost population over the same five decades.

Advances in census data compilation in Mexico and the United States beginning in 1990, and in the ability to prepare analytical maps of such data, enable a more refined description of the geographic arrangement of population in the Apache Highlands. The most recent two decennial censuses of population and housing in the United States present data in several different geographic units, including census *block groups* – areas containing 250–550 housing units and representing the second smallest geographic unit used by the US Census Bureau (US Bureau of the Census 1991). The most recent two decennial censuses of population and housing in Mexico, in turn, present data in submunicipio units called *áreas geoestadísticas básicas* (AGEBs) in 1990 and for all communities in Mexico in 2000 that can be converted to AGEBS (see Instituto Nacional de Estadística, Geografía, e Informática [INEGI] 1998b, 2002). Mapping and analyzing population (and other) data in such small units provides a more precise ability to assign geographic location to measures such as population density, revealing complex variability across the surface of the

Table 9.1 Population statistics for US counties and Mexican municipios in or adjacent to Apache Highlands Ecoregion: 1950, 1990, 2000

State	County/ Municipio	County/ Municipio area in ecoregion (%)	1950 Population	1990 Population	2000 Population	1990–2000 Average annual change (%)
Arizona	Apache[a]	–	27,767	61,591	69,423	1.2
	Cochise	99.4	31,488	97,624	117,755	1.9
	Coconino	0.4	23,910	96,591	116,320	1.9
	Gila	83.6	24,158	40,216	51,335	2.5
	Graham	73.7	12,985	26,554	33,489	2.3
	Greenlee	67.6	12,805	8,008	8,547	0.7
	Maricopa	4.5	331,770	2,122,101	3,072,149	3.8
	Mohave	9.6	8,510	93,497	155,032	5.2
	Navajo	9.6	29,446	77,658	97,470	2.3
	Pima	25.8	141,216	666,880	843,746	2.4
	Pinal	14.8	43,191	116,379	179,727	4.4
	Santa Cruz	100.0	9,344	29,676	38,381	2.6
	Yavapai	73.1	24,991	107,714	167,517	4.5
New Mexico	Catron	1.1	3,533	2,563	3,543	3.3
	Grant	20.8	21,649	27,676	31,002	1.1
	Hidalgo	95.1	5,095	5,958	5,932	–
Chihuahua	Casas Grandes	1.7	10,679	10,042	10,027	–
	Janos	61.5	4,201	10,898	10,225	−0.6
Sonora	Aconchi	60.5	1,775	2,356	2,412	0.2
	Agua Prieta	100.0	13,121	39,120	61,821	4.7
	Altar[a]	–	2,036	6,458	7,224	1.1
	Arizpe	100.0	4,659	3,855	3,397	−1.3
	Bacadéhuachi	66.1	1,659	1,499	1,347	−1.1
	Bacerac	49.7	2,573	1,775	1,369	−2.6
	Bacoachi	100.0	2,095	1,593	1,497	−0.6
	Banámichi	89.8	1,617	1,701	1,478	−1.4
	Baviácora	26.3	3,122	3,979	3,700	−0.7
	Bavispe	99.5	2,299	1,755	1,383	−2.4
	Benjamin Hill[a, b]	–	NA	5,939	5,729	−0.4
	Cananea	100.0	18,869	26,931	32,074	1.8
	Cucurpe	51.7	1,902	1,036	935	−1.0
	Cumpas	71.9	6,284	6,932	6,188	−1.1
	Divisaderos	19.8	1,098	901	823	−0.9
	Fronteras	100.0	4,183	6,336	7,872	2.2
	Granados	100.0	1,271	1,290	1,214	−0.6
	Huachinera[c]	39.7	NA	1,503	1,146	−2.7
	Huásabas	99.7	1,621	1,084	983	−1.0
	Huépac	76.8	1,236	1,262	1,144	−1.0
	Imuris	96.2	4,999	7,365	10,006	3.1
	Magdalena	57.5	9,034	20,071	24,409	2.0
	Moctezuma	31.4	3,132	3,947	4,185	0.6
	Naco	100.0	2,495	4,645	5,352	1.4
	Nácori Chico[a]	–	2,594	2,513	2,252	−1.1
	N‘ de García	100.0	5,500	13,171	14,344	0.9

(*continued*)

Table 9.1 (continued)

State	County/ Municipio	County/ Municipio area in ecoregion (%)	1950 Population	1990 Population	2000 Population	1990–2000 Average annual change (%)
	Nogales	93.5	26,016	107,936	159,103	4.0
	Opodepe	13.4	3,899	3,288	2,842	−1.4
	Rayón[a]	–	2,250	1,838	1,602	−1.4
	S.F. de Jesús[a]	–	830	470	429	−0.9
	S.M. de Horc.	21.1	4,727	2.285	5.626	9.4
	Santa Ana	16.9	9,974	12,745	13,534	0.6
	Santa Cruz	100.0	1,456	1,476	1,642	1.1
	Sáric	41.6	1,479	2,112	2,252	0.6
	Tubutama	0.6	2,186	1,842	1,790	−0.3
	Ures	5.1	8,603	10,140	9,553	−0.6
	Villa Hidalgo[d]	100.0	3,262	2,233	1,995	−1.1

Sources: Dirección General de Estadística 1952a, 1952b; INEGI (1996, 2002); US Bureau of the Census (1996, 2000)
– represents a percentage that rounds to 0; *NA* not available
[a]Outside though near ecoregion
[b]Part of Trincheras Municipio in 1950
[c]Part of Bacerac Municipio in 1950
[d]Named Oputo in 1950

Apache Highlands (Fig. 9.2).[2] Geographic precision is important in examining such data, as population density has been proposed at global (Cincotta and Engleman 2000; Gorenflo 2006; Sanderson et al. 2002) and regional scales (Brashares et al. 2001; Gorenflo 2002; Harcourt et al. 2001; Parks and Harcourt 2002) as an important indicator of human impact on biodiversity.

The map of population density by block group and AGEB indicates that ecoregion inhabitants tend to reside in definite concentrations, the hamlets, towns, and cities that characterize human settlement in most socio-cultural systems. Surrounding these communities are geographic units containing less dense population, declining with distance from the population center, again a tendency generally found in human settlement patterns. The distribution of people in the Apache Highlands Ecoregion differs from patterns found in many other places in the extremely sparse settlement found outside of communities and their immediate hinterlands – a consequence of

[2]In contrast to the 1990 census, population data by AGEB was not presented in the 2000 census for all of Mexico. To enable comparisons between the two census years, I assigned population data from the 2000 census presented by locality (INEGI 2002) to the AGEBs defined in the 1990 census (INEGI 1998b). The potential exists for introducing slight errors into the resulting geographic information coverage, through errors in digitizing AGEBs or localities – in essence causing assignment of a locality's population to the wrong AGEB. I took great care in building and reviewing the 2000 AGEB population coverage, through comparing 1990 and 2000 populations in borderline assignment decisions, though small errors likely persist.

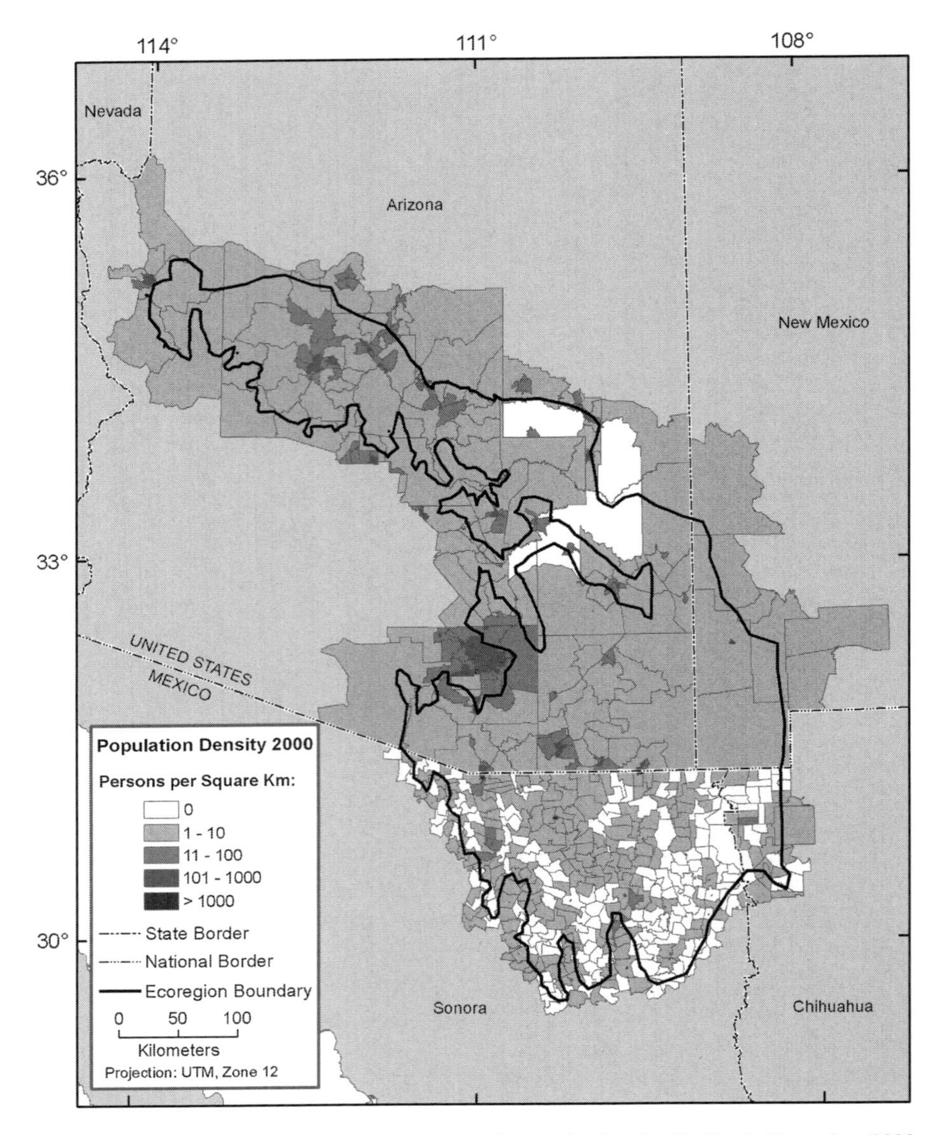

Fig. 9.2 Population density by block group and AGEB in the Apache Highlands Ecoregion, 2000 (data sources: INEGI 1998b, 2002; US Bureau of the Census 2002)

the agriculture and mining that tends to dominate rural portions of the ecoregion and the small numbers of people usually associated with such activities. Nearly all block groups in the US portion of the ecoregion contained population in 2000. In contrast, several AGEBs in the Mexico portion contained no population that same year. This possibly indicates a difference in settlement patterns – for instance, fewer dispersed ranches in Mexico. However, it likely also results from the small geographic extent of AGEBs that enables rural population to be mapped more precisely – similarly sized units north of the border possibly also would contain no human habitants in 2000. And it reveals the reliance of block group definition on resident population at a certain level, as noted in the preceding paragraph.

Recent patterns of population change also varied considerably among block groups and AGEBs in the Apache Highlands Ecoregion (Fig. 9.3). Once again, different patterns appear when one compares the US and Mexico portions of ecoregion. Consistent with evidence of widespread population growth among most counties in the region, the vast majority of block groups experienced increases in population during the 1990s. Moreover, much of the widespread population growth was quite rapid, in excess of 4.0% annually, with more than 8% of the block groups mapped minimally doubling their population (growing 7.0% or more annually) every 10 years. Some of the most rapid growth occurred near the larger communities within or immediately beyond the ecoregion boundary, the pattern around Tucson providing a good example of this tendency and the urban sprawl threatening conservation in other portions of Arizona (Gorenflo 2002). The main cause of such sprawling settlement is likely some combination of an attempt to reduce housing costs and a desire to live beyond the geographic bounds of established communities, causing land previously comprising ranches or otherwise unused to be developed for dispersed residential use.

The geographic pattern of population change during the 1990s in the Mexican portion of the ecoregion, in contrast, revealed some important differences from that found in the United States. The most obvious contrast is that slightly more than half the AGEBs in the ecoregion *lost* population between 1990 and 2000. AGEBs that experienced an increase in population over this decade often were those with established communities. The pattern seen is reminiscent of the sort often found in migration from rural settings to communities (not necessarily rural–*urban* in this case), though data do not currently exist to confirm this observation. Regardless of the underlying demographic process, as a consequence of 1990–2000 population change in the Mexico portion of the Apache Highlands Ecoregion areas of population growth tend to be more concentrated – yielding a more geographically focused pattern of population growth than found north of the border, amidst widespread rural demographic decline.

The 2000 census in both Mexico and the United States recorded individuals aged 5 years and older in 2000 by their place of residence in 1995, thereby providing insights on short-term migration. Mapping data for people in 2000 who relocated from a different county or country 5 years earlier reveals high levels of in-migration for the US portion of the ecoregion (Fig. 9.4). Minimally 15% of the population aged 5 years or more moved to their place of residence in 2000 from a different county or country over the preceding half-decade, with more than 25% in-migrating for half of the counties examined. Much less migration from another municipio or country appeared to occur in the Mexican portion of the Apache Highlands. A maximum of 17% of the resident population aged 5 years or more in-migrated from one of these places during the last half of the

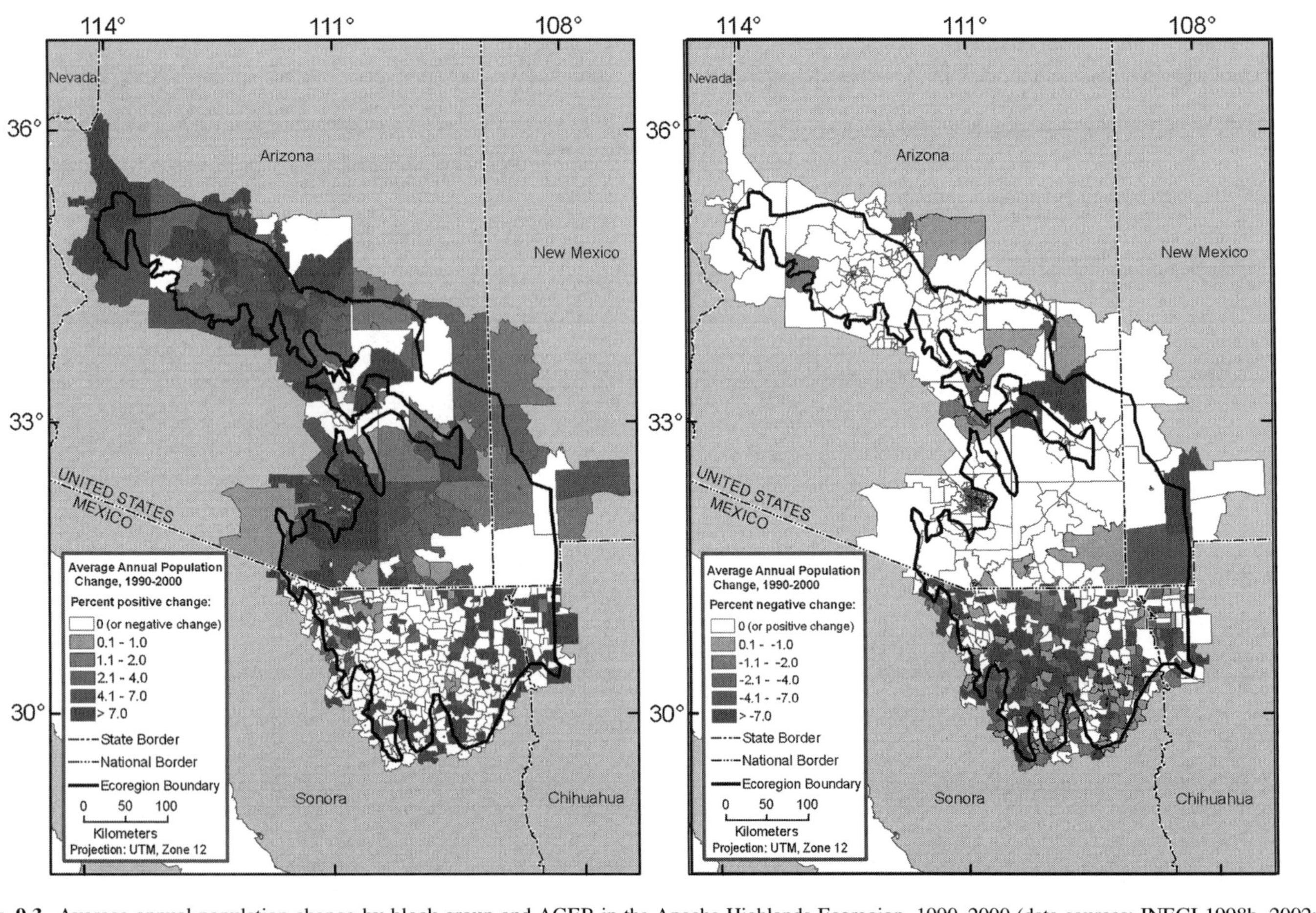

Fig. 9.3 Average annual population change by block group and AGEB in the Apache Highlands Ecoregion, 1990–2000 (data sources: INEGI 1998b, 2002; US Bureau of the Census 1992, 2002)

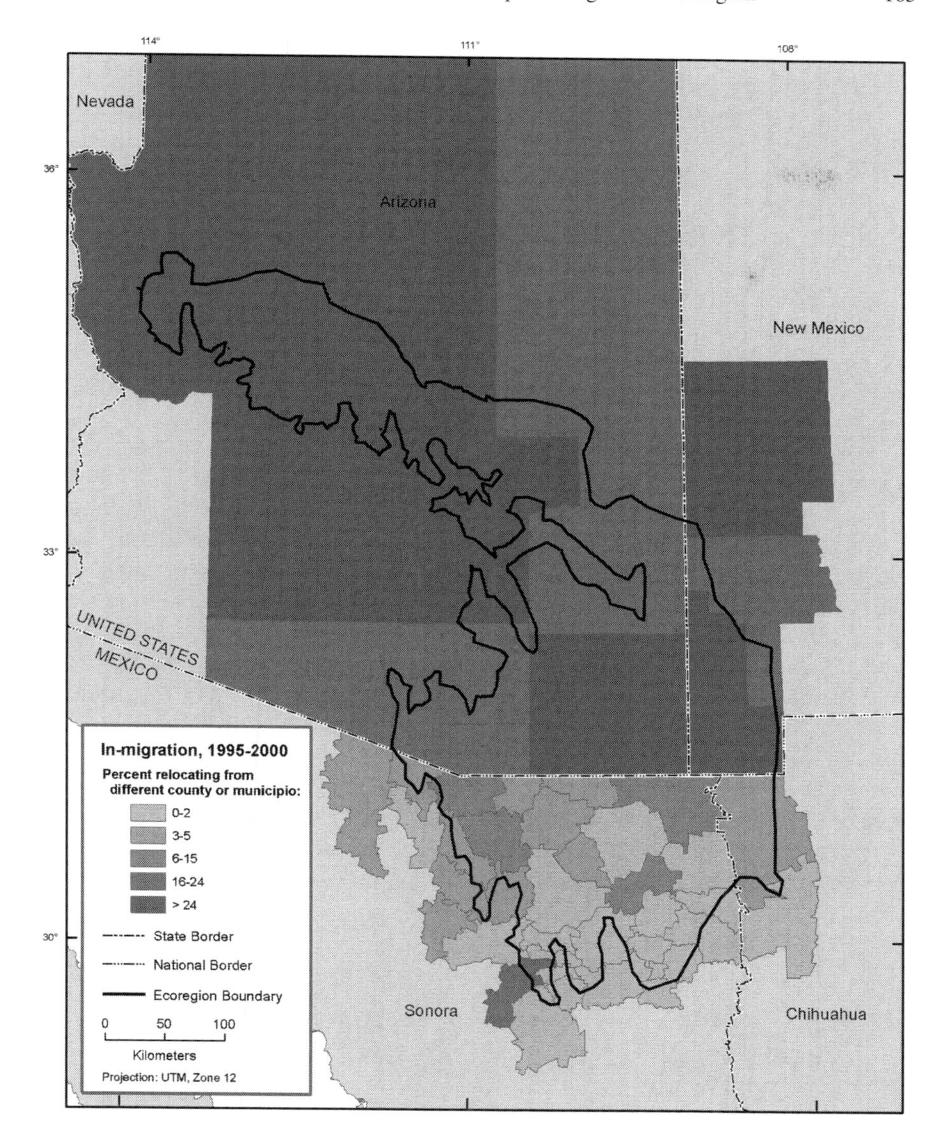

Fig. 9.4 Percent of residents aged 5 years or older in 2000 who lived in different county (in the United States portion of the ecoregion) or municipio (in the Mexico portion of the ecoregion) in 1995, by county and municipio (data sources: INEGI 2000; US Bureau of the Census 2002)

1990s to an Apache Highlands municipio, with all but 5 of the 34 municipios in the ecoregion recording rates of in-migration below 5%. Two of the exceptions were the municipios of Agua Prieta and Nogales – both adjacent to the border, containing large communities (see Fig. 9.1), and receiving relatively large numbers (as well as percentages) of in-migrants from elsewhere between 1995 and

2000. The underlying reasons for the patterns of short-term migration in each country probably both relate in part to regional and national economics. In the case of the US portion of the Apache Highlands, persisting economic growth in the southwest continued to attract large numbers of people from other parts of the country and beyond (including Mexico). In the Mexican portion of the ecoregion, economic growth in the Borderlands – along with the proximity to the US and related economic opportunities – has for decades attracted migrants from elsewhere in Mexico. However, in contrast to their counterparts north of the border, economic opportunities and other attractions to migrants in Apache Highland municipios tend to be limited to a few destinations containing relatively large communities and *maquiladoras*.

In contrast to mobility, natural increase seems to have accounted for much less population change experienced during the 1990s. Fertility among counties in the US part of the Apache Highlands part of the ecoregion in 2000 was relatively low. Crude birth rates for Arizona counties included at least partially in the Apache Highlands ranged from 10.5 to 20.8, the associated crude rate of natural increase ranging from –0.9 to 15.5 (Arizona Department of Health Services 2002). Crude birth rates for the New Mexico Counties intersected by the ecoregion ranged from 4.8 to 13.5 the same year, with the general fertility rate for those counties measuring 33.5 to 70.8 (New Mexico Department of Health 2002). In 2000, the average number of total live births to females aged 12 years or more was 2.6 for both the states of Chihuahua and Sonora (INEGI 2002). Both municipios in Chihuahua and 15 of the 34 municipios in Sonora in the ecoregion exceeded the state average number of live births, though with a maximum of 3.7 none featured fertility at a level that could drive the rapid population growth seen during the 1990s in the Apache Highlands. These low-moderate fertility rates, coupled with indications of high in-migration from other counties/municipios or other countries, indicate at best a limited role for natural increase in the population growth experienced in the ecoregion during the 1990s.

9.4 Conservation and Human Demography in the Apache Highlands Ecoregion

The maps and supporting data presented in the preceding section indicate that human population in the Apache Highlands Ecoregion tends to occur in generally localized concentrations, surrounding by nearby dispersed distributions and amidst large areas of sparse settlement. As is the case with other plant and animal species in the ecoregion, this pattern represents an adaptation to the sky island environment and intervening grasslands and riparian corridors that characterize this region. The agriculture (both crops and livestock) and mining found in this topographically dissected region do not readily support dense human population. Nevertheless, regional human demography and associated patterns of economic activity and land use have important implications for the conservation of biodiversity in this ecoregion.

In an attempt to identify conservation opportunities in the Apache Highlands Ecoregion, The Nature Conservancy Apache Highlands Ecoregional Planning Team defined 90 potential conservation sites on both sides of the border (Marshall et al. 2004). The process for defining these sites began with the identification of areas nominated by experts working in various portions of the ecoregion based on the presence of key conservation targets (flora and fauna), which served as initial nuclei for potential sites. Technical staff working on the ecoregional plan subsequently refined site locations and configurations through considering:

- Conservation target information from Natural Heritage Program databases
- Similar information from published and *gray* literature
- Evaluations of ecoregion topography, hydrography
- Examinations of land use–land cover
- Land management status

The sites resulting from this process represent the remaining locations of key conservation targets in the Apache Highlands Ecoregion, each site often containing multiple targets. In all, the conservation sites defined by the ecoregional planning team cover a large amount (nearly 42%) of the Apache Highlands. In the interest of understanding the relationship between these concentrations of remaining biodiversity and the human demography of the ecoregion, we can examine the arrangement of the former with respect to both population density and recent population change.

Figure 9.5 shows the locations of potential conservation sites in the Apache Highlands with respect to population density. In part because of the relatively large proportion of the ecoregion covered by the conservation sites, such localities occur in a range of population densities. However, a tabular summary of mean densities indicates a tendency for conservation sites to occur in sparsely populated areas, with 75 of the 90 sites occurring in areas averaging 5 persons per km^2 or less and 84 occurring in areas averaging 10 persons per km^2 or less (Table 9.2). As one would expect, sparse population is not the sole criterion yielding high biodiversity. Many sparsely settled portions of the Apache Highlands Ecoregion do not feature potential conservation sites. Such tendencies reinforce the importance of what amount to a range of ecological factors. And yet data for the Apache Highlands also suggest a role for dense population, or the activities associated with it, in possibly disrupting biodiversity. Although conservation sites do not necessarily occur in sparsely populated areas, they rarely occur in densely populated areas.

Another demographic measure that seems worthy of consideration in the context of biodiversity in a region that has witnessed considerable recent growth is population change. Figure 9.6 presents the distribution of potential conservation in the Apache Highlands Ecoregion with respect to average annual population change between 1990 and 2000, the most recent two decennial census years. Once again, the relatively large percentage of the ecoregion covered by conservation sites likely contributes to biodiversity cooccurring with a wide range of population change values (see Table 9.2) – from rapid decline to rapid growth. Broad ranges of rates of change occur within single conservation sites, particularly in the Mexican portion of the ecoregion with its combination of geographically small AGEBs,

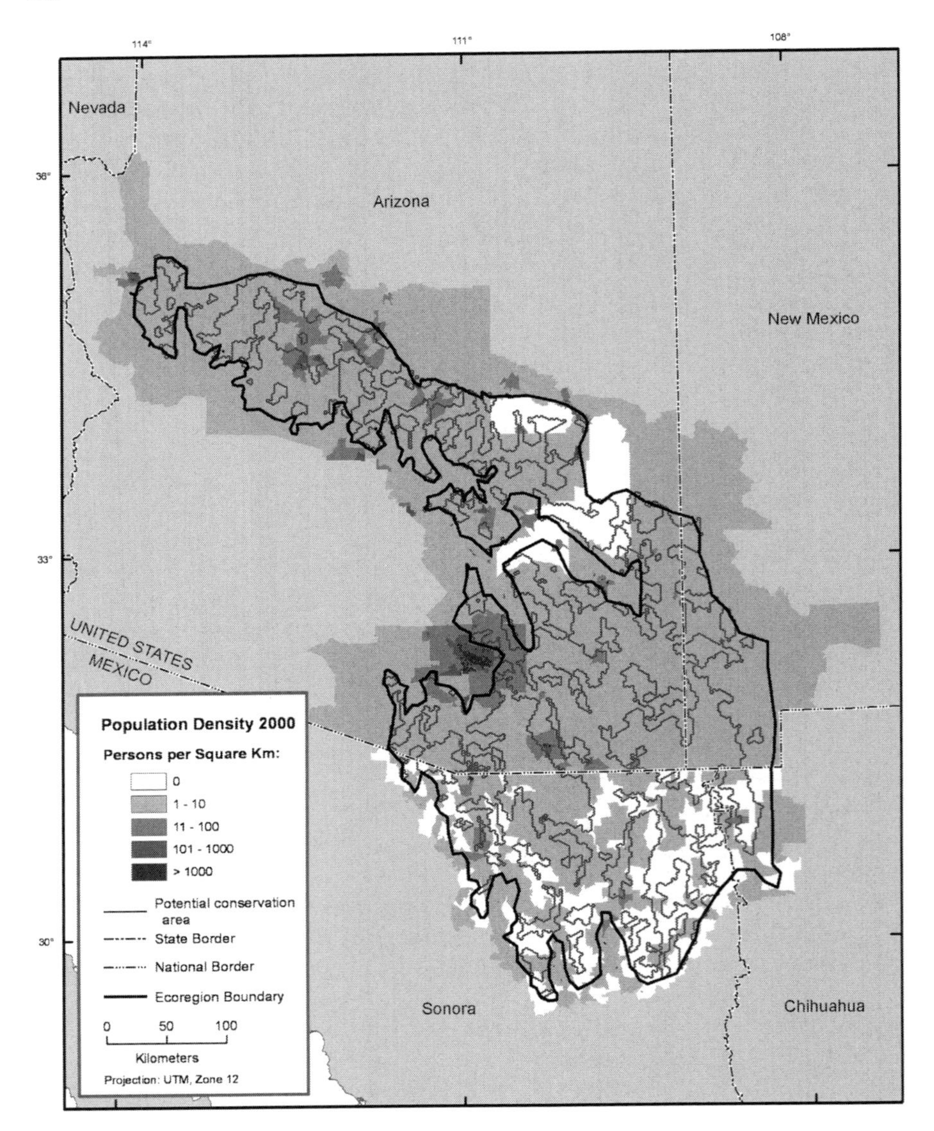

Fig. 9.5 Potential conservation sites and population density (2000) by block group and AGEB in the Apache Highlands Ecoregion (data sources: INEGI 1998b, 2002; The Nature Conservancy n.d.; US Bureau of the Census 2002)

geographically large conservation sites, and relatively rapid population shifts during the final decade of the twentieth century. Relying solely on rates of population change as an indicator of demographic activity can be misleading due to its reflection of *relative change*, with rapid population change for a small population reflecting the gain or loss of only a few people. The utility of average annual

Table 9.2 Demographic statistics for proposed conservation sites

Proposed Conservation Site Number	Persons per square kilometer			Average annual population change		
	Mean	Maximum	Standard deviation	Mean	Maximum	Standard deviation
1	0.6	0.6	0.0	10.4	10.4	0.0
2	0.5	0.6	0.2	9.6	10.4	1.7
3	0.2	0.3	0.1	5.1	5.6	0.8
4	3.6	16.4	5.4	5.8	9.3	2.2
5	0.1	0.1	0.0	5.6	5.6	0.0
6	0.7	1.3	0.4	5.3	9.3	4.2
7	0.1	0.3	0.1	6.1	7.6	0.9
8	0.8	1.3	0.3	0.5	9.3	3.0
9	16.1	872.0	55.5	6.8	14.9	3.1
10	0.2	0.2	0.0	9.6	9.6	0.0
11	7.1	174.6	12.1	11.1	14.4	4.9
12	0.8	11.4	2.5	9.5	9.6	0.4
13	5.9	8.7	4.0	5.8	9.6	2.6
14	0.6	0.6	0.0	0.3	0.3	0.0
15	2.2	6.9	1.8	5.4	5.9	0.6
16	2.7	2.7	0.4	3.7	3.7	0.2
17	0.8	0.8	0.2	2.0	2.0	0.1
18	0.4	0.8	0.2	2.5	6.6	2.7
19	5.1	6.3	2.4	4.4	4.8	0.9
20	0.3	0.8	0.2	3.0	6.5	1.4
21	2.7	10.6	3.4	4.0	9.6	1.4
22	1.7	5.1	1.3	6.6	12.8	2.4
23	5.1	5.4	0.3	0.1	3.5	4.1
24	2.2	5.1	1.9	6.7	9.3	1.6
25	2.0	5.1	2.2	8.4	12.8	3.2
26	1.2	181.2	8.2	2.5	11.8	3.9
27	0.4	0.5	0.1	3.7	6.3	1.2
28	0.4	0.5	0.2	0.9	6.3	3.3
29	0.8	1.0	0.3	4.7	11.8	6.0
30	0.8	1.0	0.3	−0.8	−0.4	0.3
31	0.5	0.5	0.0	−1.1	−1.1	0.0
32	3.7	3.7	0.0	8.6	8.6	0.0
33	3.2	4.8	1.2	7.4	8.6	2.8
34	0.2	23.2	1.3	−6.4	11.8	5.5
35	5.5	14.9	6.8	0.1	11.8	0.9
36	0.4	0.4	0.1	10.3	11.8	3.9
37	0.6	0.6	0.0	−0.5	−0.5	0.0
38	1.4	2.9	1.3	−2.5	5.0	5.9
39	1.9	1516.3	13.4	2.0	5.0	2.2
40	0.5	0.5	0.0	1.3	1.3	0.0
41	2.7	2.9	0.3	3.8	5.0	1.8
42	0.4	0.4	0.0	2.4	2.4	0.0
43	0.4	4.2	0.2	0.9	6.3	1.3
44	3.2	3763.3	60.1	2.8	4.8	1.5
46	0.5	0.5	0.0	1.3	1.3	0.0
47	0.7	2.3	0.5	1.5	2.4	0.3
48	14.2	30.2	6.6	12.9	14.6	4.0

(*continued*)

Table 9.2 (continued)

Proposed Conservation Site Number	Persons per square kilometer			Average annual population change		
	Mean	Maximum	Standard deviation	Mean	Maximum	Standard deviation
49	1.6	4.2	1.7	3.0	6.3	2.3
50	14.6	710.2	33.4	21.0	23.8	4.4
51	0.2	0.5	0.1	−9.0	1.4	1.5
52	0.5	0.5	0.0	1.4	1.5	0.6
53	4.2	34.6	6.6	4.2	14.6	2.6
54	13.4	13.4	0.0	14.6	14.6	0.0
55	1.0	1.0	0.0	0.6	0.6	0.0
56	1.7	1.7	0.0	5.2	5.2	0.0
57	8.2	8.2	0.0	4.4	4.4	0.0
58	0.6	3.0	0.3	1.5	4.0	2.2
59	2.2	3.0	0.1	3.1	3.1	0.2
60	0.8	1.0	0.2	2.0	3.3	1.4
61	1.5	7.7	2.3	4.1	9.5	2.1
62	7.9	65.2	11.1	3.0	7.9	1.9
63	1.2	8.8	1.7	2.6	12.3	7.4
64	0.1	0.1	0.0	−6.8	−6.8	0.0
65	6.6	124.2	5.6	0.3	12.3	28.3
66	5.8	2768.3	84.4	−0.6	Gained all[a]	17.0
67	0.5	68.9	1.5	3.9	Gained all[a]	31.5
68	1.1	1.2	0.3	0.0	0.7	0.4
69	0.0	0.1	0.0	−65.7	8.8	45.1
70	0.3	0.4	0.1	−10.7	−10.5	0.7
71	1.4	4.5	2.0	15.0	Gained all[a]	51.8
72	0.0	0.1	0.0	0.5	4.1	4.7
73	1.8	11.2	3.1	8.6	54.4	18.7
74	0.0	0.0	0.0	7.0	19.6	14.2
75	67.9	2941.1	411.3	3.9	23.9	6.6
76	0.1	0.2	0.0	−24.3	−7.7	31.4
77	1.9	1651.2	48.3	0.4	19.1	9.3
78	0.1	11.6	0.7	−27.5	5.8	40.6
79	0.0	2.8	0.1	−5.7	14.9	6.5
80	0.0	0.1	0.0	71.9	Gained all[a]	41.4
81	0.1	1.0	0.1	−3.7	2.1	2.8
82	1.6	1675.4	38.2	12.2	Gained all[a]	51.0
83	16.0	3921.6	228.9	11.4	Gained all[a]	41.0
84	1.2	188.3	9.0	−12.3	Gained all[a]	33.5
85	0.0	1.0	0.2	−33.8	14.9	45.1
86	0.0	0.2	0.0	8.6	24.6	6.1
87	0.0	0.2	0.0	46.4	Gained all[a]	48.3
88	0.1	0.1	0.1	26.7	Gained all[a]	48.2
89	0.1	1.3	0.3	24.8	Gained all[a]	44.1
90	0.4	4.6	0.7	12.9	Gained all[a]	32.9
91	0.0	0.1	0.0	−13.3	0.0	4.6

[a]Refers to a geographic unit that had 0 inhabitants in 1990 and one or more in 2000; in such cases, average annual change cannot be calculated as the starting number is 0

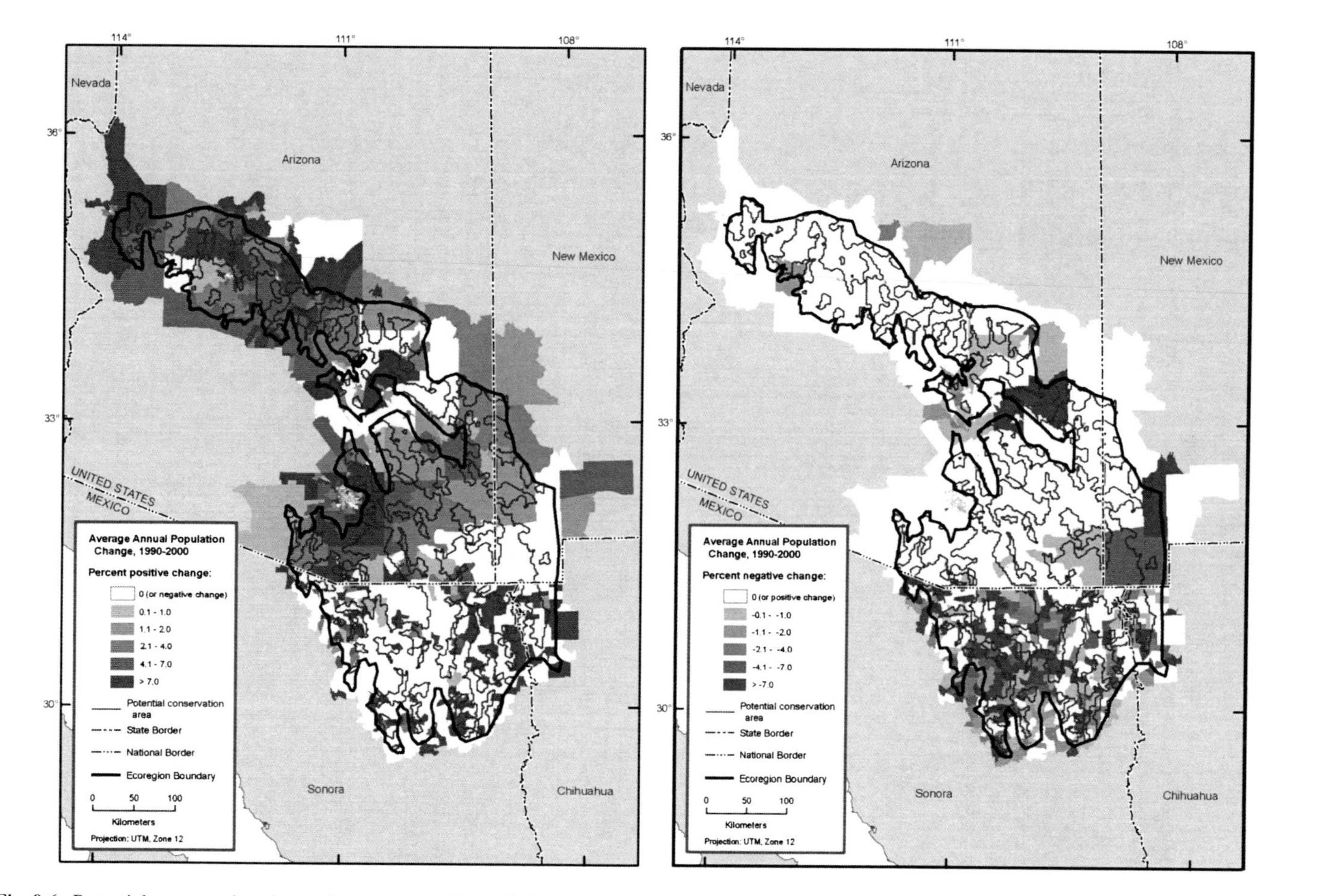

Fig. 9.6 Potential conservation sites and average annual population change (1990–2000) by block group and AGEB in the Apache Highlands Ecoregion (data sources: INEGI 1998b, 2002; The Nature Conservancy n.d.; US Bureau of the Census 1992, 2002)

population change as an indicator of places incompatible with biodiversity is enhanced greatly by focusing on those localities (block groups and AGEBs, in this case) with comparatively high population densities, as discussed below.

The utility of human population data or some variant of them as a predictor of biodiversity is complicated by two key factors in the Apache Highlands Ecoregion: the relationship of population to various human activities and the role of geographically indirect impacts. The first acknowledges that many types of human activity completely incompatible with conservation can occur in a range of population density settings. For instance, in the case of the US portions of the Apache Highlands satellite imagery from 1992 indicates that agriculture occurs in areas with population densities as low as 0.5 persons per km^2 (US Geological Survey and Environmental Protection Agency 2000a, b).[3] Similar evidence exists for the Mexican portion of the ecoregion, using statistical data as evidence of economic activity (INEGI 1998a, b). Although the presence of people arranged in such sparse settlement would not itself lead to severe habitat destruction – as a consequence of dense housing and other human-related infrastructure – the *activities* in which they are engaged might (and often do) compromise biodiversity. Human population, primarily represented as density, generally serves as a surrogate of varying utility for such activities.

The issue of geographically indirect impacts is one of people living in a particular locality and causing impacts in another. Agriculture is one example of such impacts, where people engaged in agriculture (or another activity generally incompatible with biodiversity) in one area often reside apart from their fields. For arid regions in general, and the Apache Highlands in particular, such indirect impacts often involve surface and subsurface hydrology, affecting the riparian areas so vital to the ecological health of such geographic settings (see Kouros 1998). One of the best known, most important, and most highly threatened riparian areas is the San Pedro River, which lies in the eastern part of the Apache Highlands (Council for Environmental Cooperation 1999). Providing essential habitat to nearly 400 species of migratory birds, more than 100 species of butterflies, more than 80 species of mammals, and nearly 50 species of amphibians and reptiles, the San Pedro riparian area is particularly important to the 1–4 million birds that stop there during their annual migrations between Latin America and points north (Hanson 2001). The extraction of ground water for domestic use, particularly by the US Army installation of Ft. Huachuca and the nearby community of Sierra Vista, and for agricultural irrigation interrupts the hydrologic cycle that recharges the regional aquifer. Effects are both local near the demand points and more distant in other parts of the compromised riparian system. The future of this key component of the Apache Highlands remains uncertain, depending largely on population

[3]For the sake of the argument presented here, I used available interpreted Landsat imagery despite its dating to 1992. Because of the date of the imagery, I used population density data from 1990 to conduct this phase of the analysis – closer to the date of the imagery than more recent census data.

growth, per capita domestic water use, and local land use which all interact through the surface and subsurface hydrology (Steinitz et al. 2003).

What, then, can the study of human population tell us of biodiversity and potential biodiversity conservation in the Apache Highlands? As noted, it is important to recognize that impacts to biodiversity are not generated by the mere presence of people, but rather by their activities. Demographic data in some cases can be used to identify locations of activities incompatible with biodiversity. High population density (for instance, more than 200 persons per km^2) indicates a level of human activity that can indeed disrupt biodiversity – such as residential behavior associated with large amounts of infrastructure (residential construction, commercial construction, transportation networks, etc.) that itself displaces natural habitat. Of the 90 sites identified based on remaining biodiversity, none was associated with such dense human habitation. Only one site was associated with mean population densities greater than 50 persons per km^2 – Site 75 (see Table 9.2), an elongated site which coincides in part with the community of Imuris, Sonora, and its population density of more than 67.9 persons per km^2. All but 5 of the remaining 89 sites had mean population densities less than 10 persons per km^2. Average annual change is much less effective in the Apache Highlands at identifying a level beyond which biodiversity seems to be compromised. For instance, of the 90 potential conservation sites, 23 occurred in areas with mean population growth of 7.0% or more annually – that is, doubling every decade – which by most standards is quite rapid.

Based on data from the Apache Highlands, a population density 10 persons per km^2 tends to serve as a rough threshold of sorts, beyond which noteworthy biodiversity is unlikely. Proposed conservation sites that occur in locations that will exceed this population density in coming years – calculated through applying 1990–2000 growth rates to 2000 population data – may face conservation risks in the near future (Fig. 9.7). Greater confidence awaits projections of likely future land use patterns for the Apache Highlands and an improved understanding of the relationship between various forms of land use and key processes upon which various conservation targets rely.

9.5 Concluding Remarks

The Apache Highlands comprises an ecoregion partially lying in the US–Mexico Borderlands and containing considerable remaining biodiversity amidst a dispersed, though growing, human population. This study has focused on the demographics of this region – notably the density of human habitation, the nature of recent population change, and causes of this change. Through employing geographic information system technology and mapping population-related variables in small geographic units, it is possible to propose certain relationships between the demographics of humans and the diversity of other species in the region.

The nature of human demography in the Apache Highlands Ecoregion has historically been one of relatively sparse settlement. This certainly was the case

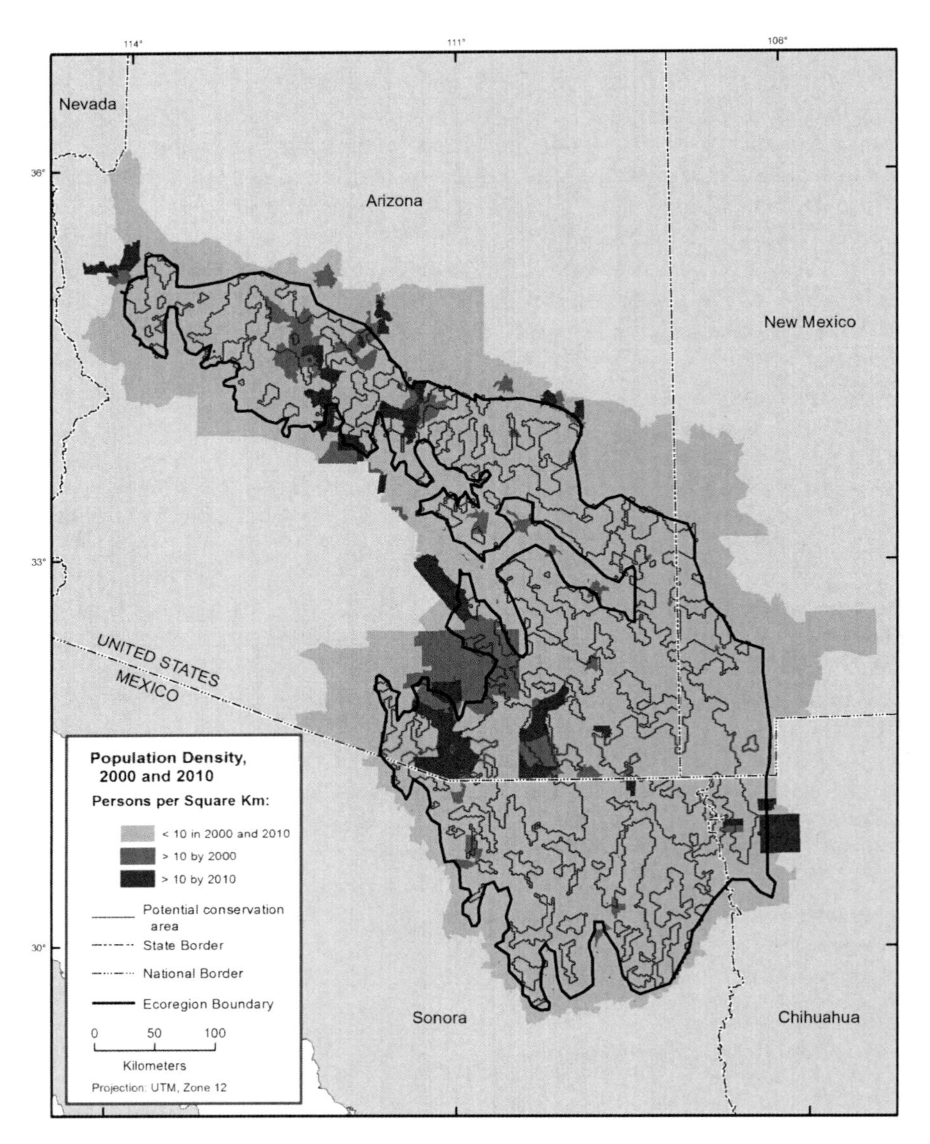

Fig. 9.7 Block groups and AGEBs in the Apache Highlands Ecoregion exceeding 10 persons per km^2 in 2000 and projected to exceed 10 persons per km^2 in 2010

during prehistoric times and most historic times. Although population in parts of the ecoregion grew during the second half of the twentieth century, settlement remained relatively sparse – with concentrations of population scattered about on both sides of the border in the form of communities and their hinterlands. Coupled with historically limited potential for the activities that tend to compromise biodiversity – such as large-scale agriculture and commercial forest harvest – the situation that one faces is a few instances of concentrated human presence and limited activities that cause biodiversity loss. In all likelihood, as a result of these human settlement and activity patterns in the Apache Highlands, much of the ecoregion lies within potential conservation sites.

Available data on population and potential conservation sites indicate that dense populations generally are incompatible with biodiversity. This seemingly obvious conclusion is enhanced by the sense that biodiversity begins to fall off at densities of about 10 persons per km^2 – a conclusion that is ecoregion-specific with regard both to the nature of human habitation and ecology of this particular area. Lower population densities do not necessarily mark areas of high biodiversity, in part because the potential for destructive activities exists in areas with relatively few people and in part because biodiversity is not uniformly rich across the entire ecoregion.

Given documented demographic trends in the Apache Highlands Ecoregion, what does the future hold with respect to biodiversity? Population likely will remain concentrated primarily in communities, where the highest densities will occur, around which settlement density will decline but often be present. The marked population growth between 1990 and 2000 will continue, barring major economic shifts, leading to increased dispersal around existing population centers. Those areas near current settlements with potential conservation sites may well change to localities incompatible with biodiversity. Further disruption is possible through the growing impact that increased human habitation will have on water and the network of riparian areas that are so crucial to ecosystem function throughout the Apache Highlands. Localities protected from development, largely occurring through public ownership on both sides of the border, can help limit direct impacts of population growth and development, but have limited effect on overuse of key regional resources such as water.

Humans increasingly form the dominant species in many terrestrial ecosystems, with their potential for either exploiting other species or the habitat upon which they rely having a massive impact on biodiversity. As a result, it becomes increasingly important to consider human impacts in the context of conservation planning. The most sensitive conservation settings are precisely the type considered in this study – an ecoregion with considerable remaining biodiversity, a particularly fragile natural environment, and rapid development. In the Apache Highlands Ecoregion, areas of high population density are scattered and generally geographically concentrated. Biodiversity does not occur in these places, and as a consequence examining population appears to be a useful avenue for identifying those areas which do not support (and will not support) noteworthy biodiversity. For the remaining areas with less human population, the utility of using population as an indicator of remaining biodiversity declines considerably, and it is here that one must examine complementary data. Considering such data, identifying the demographic *and* land use conditions inconsistent with maintaining biodiversity, and engaging decision makers who can influence the pattern of settlement, land use, and development, will greatly enhance our ability to conserve biodiversity in the long term.

Acknowledgments Rob Marshall and Anne Gondor of The Nature Conservancy's Arizona Chapter kindly provided data on conservation sites essential for the preceding study. The latter also read and commented on an early version of this paper, as did Katrina Brandon. Joe Adduci helped to clean up one of the geographic information system coverages used to produce the maps presented in the paper, and recommendations by Mark Denil greatly improved map design. With characteristic patience and understanding, my wife, Grace, and our daughters, Maria, Annie, and Lauren, went well beyond the call of duty to provide me with additional time to assemble this study amidst a particularly busy schedule.

References

Alegría T (1992) Desarollo urbano en la frontera México-Estados Unidos. Consejo Nacional para la Cultura y las Artes, México, DF

Arizona Department of Health Services (2002) Arizona health status and vital statistics, 2001 report. Arizona Department of Health Statistics, Phoenix, AZ

Basso KH (1983) Western Apache. In: Ortiz A (ed) Handbook of North American Indians, vol 10. Smithsonian Institution Press, Washington, DC, pp 462–488

Brashares JS, Arcese P, Sam MK (2001) Human demography and reserve size predict wildlife extinction in West Africa. Proc R Soc Lond Ser B 268:2473–2478

Canales A (1999) The new face of urbanization in Mexico. Borderlines 7(7):1–4

Cincotta RP, Engleman R (2000) Nature's place: human population density and the future of biological diversity. Population Action International, Washington, DC

Cordell L (1997) Archaeology of the Southwest. Academic, New York, NY

Council for Environmental Cooperation (1999) Ribbon of life. An Agenda for preserving transboundary migratory bird habitat on the upper San Pedro River. Council for Environmental Cooperation, Montreal, Quebec

Dirección General de Estadística (1952a) Séptimo censo general de población, 6 de junio de 1950, estado de Chihuahua. Secretaria de Economía, México, DF

Dirección General de Estadística (1952b) Séptimo censo general de población, 6 de junio de 1950, estado de Sonora. Secretaria de Economía, México, DF

Dunnigan T (1983) Lower Pima. In: Ortiz A (ed) Handbook of North American Indians, vol 10. Smithsonian Institution Press, Washington, DC, pp 217–229

Environmental Systems Research Institute (1992) Digital Chart of the World. Environmental Systems Research Institute, Redlands, CA (Data available at http://www.maproom.psu.edu/dcw/dcw_about.shtml)

Ezell PH (1983) History of the Pima. In: Ortiz A (ed) Handbook of North American Indians, vol 10. Smithsonian Institution Press, Washington, DC, pp 149–160

Gerhard P (1993) The north frontier of New Spain, Revised edn. University of Oklahoma Press, Norman, OK

Gifford EW (1936) Northwestern and Western Yavapai. Univ Calif Publ Am Archaeol Ethnol 34:247–354

Goodwin G (1942) The social organization of the Western Apache. University of Chicago Press, Chicago, IL

Gorenflo LJ (2002) The evaluation of human population in conservation planning: an example from the Sonoran Desert Ecoregion. Publications for Capacity Building. The Nature Conservancy, Arlington, VA

Gorenflo LJ (2006) Population. In: Sanderson EW, Robles Gil P, Mittermeier CG, Martin VG, Kormos CF (eds) The human footprint: challenges for wilderness and biodiversity. CEMEX, Mexico, DF, pp 63–67

Griffen WB (1969) Culture change and shifting populations in central northern Mexico. Anthropological Papers of the University of Arizona 13, Tucson, AZ

Hackenberg RA (1983) Pima and Papago ecological adaptations. In: Ortiz A (ed) Handbook of North American Indians, vol 10. Smithsonian Institution Press, Washington, DC, pp 161–177

Hanson RB (2001) The San Pedro River. A discovery guide. University of Arizona Press, Tucson, AZ

Harcourt AH, Parks SA, Woodroffe R (2001) Human density as an influence on species/area relationships: double jeopardy for small African reserves? Biodivers Conserv 10:1011–1026

Hartmann WK (1989) Desert heart: chronicles of the Sonoran Desert. Fisher Books, Tucson

Haury EW (1936) The Mogollon culture of southwestern New Mexico. Medallion Papers 20. Gila Pueblo, Globe, AZ

Haury EW (1985) Mogollon culture in the Forestdale Valley, east-central Arizona. Univerisity of Arizona Press, Tucson, AZ

Hinton TB (1983) Southern periphery: West. In: Ortiz A (ed) Handbook of North American Indians, vol 10. Smithsonian Institution Press, Washington, DC, pp 315–328

Instituto Nacional de Estadística, Geografía, e Informática (1996) Los municipios de México, información censal (Compact disk). Instituto National de Estadística, Geografía, e Informática, Aguascalientes

Instituto Nacional de Estadística, Geografía, e Informática (1998a) AGROS: Información Censal Agropecuaria (Compact disks (2)). Instituto National de Estadística, Geografía, e Informática, Aguascalientes

Instituto Nacional de Estadística, Geografía, e Informática (1998b) Niveles de bienestar por Ageb (Compact disk). Instituto National de Estadística, Geografia, e Informática, Aguascalientes

Instituto Nacional de Estadística, Geografía, e Informática (2000) XII Censo general de población y vivienda Sintesis de Resultatados; 2001 (accessed for the states of Chihuahua and Sonora, respectively, via http://hades.inegi.gob.mx/sitio_inegi/difusion/espanol/poblacion/definitivos/chih/sintesis/indice.html and http://hades.inegi.gob.mx/sitio_inegi/difusion/espanol/poblacion/definitivos/son/sintesis/indice.html)

Instituto Nacional de Estadística, Geografía, e Informática (2002) Localidades de la República Mexicana. Instituto National de Estadística, Geografía, e Informática, Aguascalientes [Editado por Comisión Nacional para el Conocimiento y Uso de la Biodiversidad]

Jackson RH (1998) Northern New Spain. In: Jackson RH (ed) New views of Borderlands history. University of New Mexico Press, Albuquerque, NM, pp 73–106

Kouros G (1998) Borderlands biodiversity: walking a thin line. Borderlines 6(2):1–6

Lorey D (ed) (1990) United States–Mexico border statistics since 1900. UCLA Latin American Center Publications, Los Angeles, CA

Lorey D (ed) (1993) United States–Mexico border statistics since 1900: 1990 update. UCLA Latin American Center Publications, Los Angeles, CA

Martin PS (1979) Mogollon. In: Ortiz A (ed) Handbook of North American Indians, vol 9. Smithsonian Institution Press, Washington, DC, pp 61–74

Marshall RM et al (2004) An ecological analysis of conservation priorities in the Apache Highlands Ecoregion. Arizona State Program, The Nature Conservancy, Tucson, AZ

National Aeronautics and Space Administration (2004) Shuttle Radar Topography Mission. National Aeronautics and Space Administration, Washington, DC (Data available at http://www2.jpl.nasa.gov/srtm/cbanddataproducts.html)

National Heritage Institute (1998) Environmental degradation and migration: the US–Mexico case study. Environ Change Security Project Rep 4:61–67

New Mexico Department of Health (2002) New Mexico county health profiles. Office of New Mexico Vital Records and Health Statistics, Public Health Division, New Mexico Department of Health, Albuquerque, NM

Opler ME (1941) An Apache life-way: the economic, social, and religious institutions of the Chiricahua Indians. University of Chicago Press, Chicago, IL

Ortega NS (1993) Un ensayo de historia regional: El noroeste de México 1530–1880. Universidad Nacional Autónoma de México, México, DF

Ortiz A (ed) (1983) The Southwest. Handbook of North American Indians, vol 10. Smithsonian Institution Press, Washington, DC
Parks SA, Harcourt AH (2002) Reserve size, local human density, and mammalian extinctions in US protected areas. Conserv Biol 16:800–808
Peach J, Williams J (2000) Population and economic dynamics on the US–Mexico border: past, present, and future. In: Ganster P (ed) The US–Mexico Border: a road map for a sustainable 2020. Southwest Center for Environmental Research and Policy, San Diego, pp 37–72
Pennington CW (1980) The Pima Bajo of Central Sonora, Mexico (2 vols). University of Utah Press, Salt Lake City
Sable MH (1989) Las maquiladoras: assembly and manufacturing plants on the United States-Mexico border. Haworth, New York
Sanderson SE (1981) Agrarian populism and the Mexican state: the struggle for land in Sonora. University of California Press, Berkeley, CA
Sanderson EW, Jaiteh M, Levy MA, Redford KH, Wannebo AV, Woolmer G (2002) The human footprint and the last of the wild. BioScience 52:891–904
Sauer CO (1934) The distribution of aboriginal tribes and languages in northwestern Mexico. Ibero-Americana 9. University of California Press, Berkeley and Los Angeles, CA
Sauer CO (1935) Aboriginal population of northwestern Mexico. Ibero-Americana 10. University of California Press, Berkeley and Los Angeles, CA
Sheridan TC (1995) Arizona: a history. University of Arizona Press, Tucson, AZ
Spicer EH (1962) Cycles of conquest: the impact of Spain, Mexico, and the United States on the Indians of the Southwest, 1533-1960. University of Arizona Press, Tucson, AZ
Steinitz C et al (2003) Alternative futures for changing landscapes. The upper San Pedro River Basin in Arizona and Sonora. Island, Washington, DC
The Nature Conservancy (n.d.) Geographic information system data on conservation sites provided by Arizona State Program, Tucson, AZ
The Nature Conservancy (2000) Conservation by design. The Nature Conservancy, Arlington, VA
US Bureau of the Census (1991) 1990 census of population and housing: summary of population and housing characteristics, Arizona. US Bureau of the Census, Washington, DC
US Bureau of the Census (1992) 1990 census of population and housing, summary file 3A (Compact disk). US Bureau of the Census, Washington, DC
US Bureau of the Census (1996) Population of states and counties of the United States: 1790 to 1990. US Bureau of the Census, Washington, DC
US Bureau of the Census (2000) 2000 census of population and housing, state and county statistics. US Bureau of the Census, Washington, DC (Statistics for Arizona available at http://quickfacts.census.gov/qfd/states/04000.html and for New Mexico at http://quickfacts.census.gov/qfd/states/35000.html)
US Bureau of the Census (2002) Census 2000 Summary File 3 (Compact disk). US Bureau of the Census, Washington, DC
US Geological Survey and US Environmental Protection Agency (2000a) National land cover data, Arizona. EROS Data Center, Multi-resolution Regional Land Cover Characterization Project, Sioux Falls, ID
US Geological Survey and US Environmental Protection Agency (2000b) National land cover data, New Mexico. EROS Data Center, Multi-resolution Regional Land Cover Characterization Project, Sioux Falls, ID
West RC (1993) Sonora, its geographical personality. University of Texas Press, Austin, TX

Chapter 10
Long-Term Ecological Effects of Demographic and Socioeconomic Factors in Wolong Nature Reserve (China)

Li An, Marc Linderman, Guangming He, Zhiyun Ouyang, and Jianguo Liu

10.1 Introduction

Human population has exerted enormous impacts on biodiversity, even in areas with "biodiversity hotspots" identified by Myers et al. (2000). For instance, the population density in 1995 and the population growth rate between 1995 and 2000 in biodiversity hotspots were substantially higher than world averages, suggesting a high risk of habitat degradation and species extinction (Cincotta et al. 2000). Many regression models have been built to establish correlated relationships between biodiversity and population (e.g., Forester and Machlis 1996; Brashares et al. 2001; Veech 2003; McKee et al. 2004). These models are important and necessary, but they use aggregate variables such as population size, density, and growth rate, which may mask the underlying mechanisms of biodiversity loss and could result in potentially misleading conclusions. For example, does a declining population growth reduce the impact on biodiversity? Although global population growth has been slowing down, household growth has been much faster than population growth (Liu et al. 2003). The continued reduction in household size (i.e., number of people in a household) has contributed substantially to the rapid increase in household numbers across the world, particularly in countries with biodiversity hotspots. Even in areas with a declining population size, there has nevertheless been a substantial increase in the number of households (Liu et al. 2003). More households require more land and construction materials and generate more waste. Furthermore, smaller households use energy and other resources less efficiently on a per capita basis (Liu et al. 2003). Thus, impacts on biodiversity may be increased despite a decline in population growth.

To uncover the mechanisms associated with human population that underlie biodiversity loss and provide valuable information for biodiversity conservation, it is crucial to go beyond regression analyses and examine how demographic (e.g., population processes and distribution) and socioeconomic factors affect biodiversity at the landscape level. As many effects may not surface over a short period of time, it is essential to conduct long-term studies. However, landscape level long-term studies are costly, and it is very difficult to conduct experiments on some types of subjects, such as people. Fortunately, systems modeling has become a useful tool

R.P. Cincotta and L.J. Gorenflo (eds.), *Human Population: Its Influences on Biological Diversity*, Ecological Studies 214,
DOI 10.1007/978-3-642-16707-2_10,

to facilitate landscape-scale long-term simulation experiments (Liu and Taylor 2002). For this chapter, we applied a systems model we had developed (An et al. 2005) to study the long-term ecological effects of demographic and socioeconomic factors in Wolong Nature Reserve, southwestern China.

10.2 Profile of Wolong Nature Reserve

Wolong Nature Reserve (Fig. 10.1) is located in Sichuan Province, one of China's most populated provinces. Designated in 1975 with a total area of approximately 2,000 km^2 to conserve the endangered giant panda, it is characterized by a dramatically varying biophysical environment. With elevations ranging from approximately 1,200 m to over 6,200 m, it encompasses several climatic zones and contains over 6,000 plant, insect, and animal species. Among them, 60 are on the national protection list (Tan et al. 1995). For such reasons, Wolong and its adjacent regions were listed as one of the 25 global "biodiversity hotspots" defined in the late 1990s (Myers et al. 2000), and in Conservation International's more recently expanded set of 34 biodiversity hotspots. Roughly 60% of Wolong is situated within the Mountains of Southwest China Hotspot (Mittermeier et al. 2005). Wolong enjoys high domestic standing as a "flagship" reserve and receives considerable domestic and international financial and technical support (Liu et al. 2001).

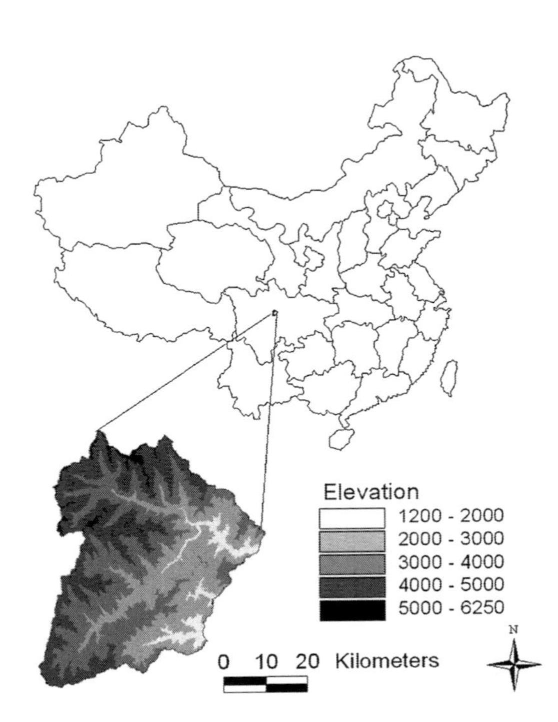

Fig. 10.1 The location and elevation of Wolong Nature Reserve in China (An et al. 2005)

In 2000, approximately 4,413 farmers lived in the reserve, mostly along the sides of the two main rivers running through the reserve; this population is made up of four ethnic groups: Han, Tibetan, Qiang, and Hui (Liu et al. 1999a). Local residents cut wood in the forests (on which pandas depend for habitat) for cooking and heating their households in winter, using electricity mainly for lighting and electronic appliances. Only a small portion of the households uses electricity for cooking and heating (An et al. 2001, 2002). No local market exists for fuelwood transaction, and the farmers collect fuelwood primarily in winter for their own use in the following year. Spending enormous amounts of time and energy for fuelwood collection, local residents find it increasingly difficult to collect fuelwood due to the shrinking forest area and topography characterized by high mountains and deep valleys. The reserve administration has implemented many policies to restrict fuelwood collection, including banning fuelwood collection in key habitat areas and prohibiting some tree species from being harvested. Enforcement of these fuelwood restriction policies tends to be ineffective because forests are a common property and difficult to monitor, given the rugged terrain. Even though electricity was available in the reserve, there was a continued increase in annual fuelwood consumption (from 4,000 to 10,000 m^3 from 1975 to 1999), contributing to a reduction of over 20,000 ha of panda habitat (Liu et al. 1999b). Degradation of forests and panda habitat was undoubtedly a factor in the reported decrease in panda population, i.e., from 145 individuals in 1974 (Schaller et al. 1985) to 72 in 1986 (China's Ministry of Forestry and World Wildlife Fund 1989).

The serious threat to the giant pandas comes from the subsistence needs of a fast growing population experiencing dramatic changes in age structure and other aspects. An even faster growth in household numbers may be contributing to the threat as well. Human population increased by 72.4% (from 2,560 in 1975 to 4,413 in 2000, an average 2.9% per year); but the number of households increased by 129.9% over the same period (from 421 to 968, an average of 5.2% per year; Fig. 10.2a).

The rapid increase in human population is due to a low mortality rate coupled with a higher fertility rate relative to other areas of China. Because of China's restrictions on migration through the household registration system (known as Hukou), along with Wolong's special standing as a nature reserve, the only legal way for people outside the reserve to migrate into the reserve is through marriage, and the number of such migrants is relatively low (An et al. 2001). For instance, 49 people (9 males and 40 females) migrated into Wolong through marriage over the period between 1996 and 2000 (An et al. 2005). The mortality rate has declined over decades in China (Fang 1993), and it is probable that Wolong has experienced the same trend (we do not have longitudinal data about mortality in Wolong). However, from the perspective of fertility, China's "one-child policy" has not been applied to minority ethnic groups such as Tibetans, which constitute over 75% of the total population in Wolong. This explains the relatively high total fertility rate (TFR) in Wolong, which was 2.5 in the 1990s (Liu et al. 1999a). Our field observations show that the fertility rate prior to 1990 likely exceeded 2.5, which contributed to the overall annual growth rate of 2.9% from 1975 to 2000,

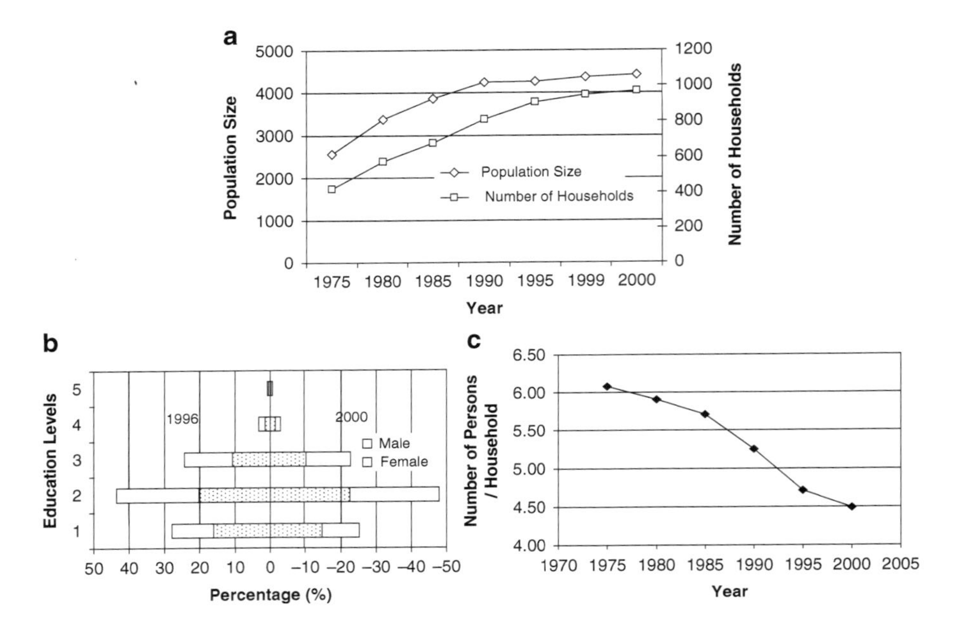

Fig. 10.2 Population of Wolong Nature Reserve. (**a**) The dynamics of population size and the number of households in Wolong between 1975 and 2000. (**b**) Education levels of Wolong population in 1996 and 2000, where level 1 is for illiteracy, 2 for elementary school, 3 for middle school, 4 for high school, and 5 for college, technical school, or higher. (**c**) Changes in household size from 1975 to 2000

together with other influences, such as a higher proportion of young people reaching the age of fertility during this period.

Previous research has shown that a change in the age structure could have a significant impact on biodiversity: the more young adults living in Wolong, the more forest may be cut down (Liu et al. 1999a). Average ages of local residents increased from 1982 to 1996, with a decreased portion of the people belonging to the 0–4, 5–9, and 10–14 age groups (Liu et al. 1999a; Wolong Nature Reserve 1997, 2000, Fig. 10.3). Changes between the 1996 and 2000 age structures were not as obvious as those between 1982 and 1996, probably because of the shorter time period. Overall, the groups that constitute the labor force (20–59 years) dominated the local population, consistent with China's general trend characterized by a decreased proportion of children (0–14 years) and an increased proportion of working-age (15–64 years) individuals (Hussain 2002). This decline was partly due to the "later, longer, and fewer" (wan xi shao) family planning campaign, encouraging or requiring couples to bear children later in life (later), prolonging the time between the births of two consecutive children if more than one child is allowed (longer), and having as few children as possible (fewer), which later developed into the more strict "one-child policy" in most parts of China, especially in cities (Feng and Hao 1992).

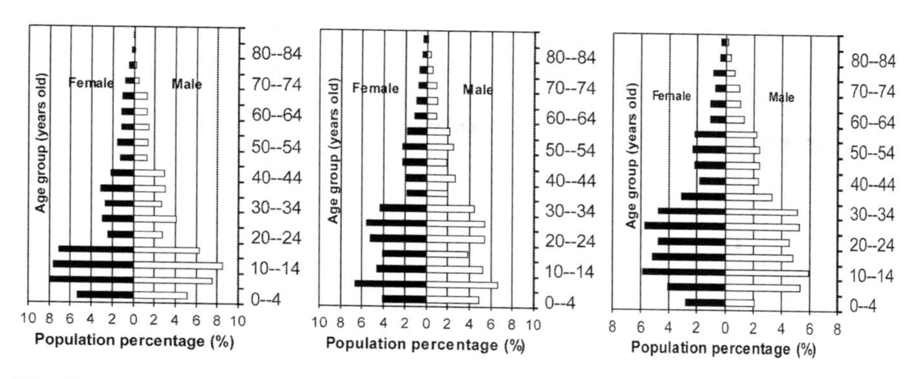

Fig. 10.3 Age and sex structures of Wolong population in 1982, 1996, and 2000

Young people's increasing preference for living independently may partially explain the faster increase in the number of households in Wolong. Traditionally, Chinese people were accustomed to a lifestyle of many generations under one roof (Liu et al. 2001); in rural areas of China, the patrilineal extended family is still the prevailing order, and the majority of the elderly people tend to live with their children (with sons in particular; Cooney and Shi 1999). Our research results, however, have shown that although young adults in Wolong care about the adverse effects associated with leaving their parental home (such as responsibility for housework and taking care of young children), many of them prefer to live independently as long as resources (land and timber in particular) allow them to do so (An et al. 2003). The proportions of larger households (6 or more people/household) declined from 1996 to 2000, while those with smaller households grew, with one exception: the proportion of households with three people declined slightly. Overall, the average household size declined from 6.08 in 1975 (Liu et al. unpublished data) to 4.60 in 1996 to 4.45 in 2000 (An et al. unpublished data) (Fig. 10.2c).

Temporary migration has become a hot topic in today's China, because seasonal workers in cities who maintain their permanent residence (characterized by the Hukou System) in rural areas have affected nearly all aspects of China's economy (Ma 1999). Wolong has seen a relatively lower proportion of such migration, probably for several reasons. First, its special standing as a nature reserve has provided subsidies (e.g., lower agricultural tax) for local people that are unavailable to those living elsewhere, and its local ecotourism centered on watching the panda in the breeding center has attracted some local people to work in local businesses such as hotels and restaurants. Our field observations have also shown that some young people work outside the reserve, returning only for holidays, such as the spring festival. As the gaps between Wolong and outside areas (Wolong vs. wealthier urban areas and Wolong vs. nearby poorer rural areas in terms of economic growth and job/education opportunities) widen, migration through marriages is expected to increase substantially. Therefore, we focus our concerns on migration through marriages, despite their relatively low numbers in the recent past.[1]

Females had lower education levels than males: in both 1996 and 2000, a higher proportion of females were at the illiterate level, and a lower proportion of females belonged to each of the other levels (Fig. 10.2b). This suggests that girls did not have the same chances for education as boys, probably due to the traditional patrilineal extended family structure. The gender difference in education may increase the probability that girls migrate out of Wolong through approaches other than education-migration (i.e., moving out of Wolong through going to college and finding jobs outside).

The pooled data (data for both females and males) show that the overall education situation improved over time: illiteracy declined from 31% in 1982 (Liu et al. 1999a) to 28% in 1996 and to 25% in 2000. This change may indicate that in the future, a higher proportion of children may pass the national college entrance examinations, go to college, and settle down in cities after finding jobs there, which could be a source of family pride for most of the parents in Wolong. According to Liu et al. (2001), the vast majority of middle-aged and elderly residents were not willing to move out of the reserve due to their low level of educational attainment (Fig. 10.2b), difficulties in finding jobs in cities, and/or difficulties in adapting to outside environment. However, they generally took pride in their children and grandchildren doing so.

All such demographic and socioeconomic factors may affect panda habitat to varying degrees, especially over a long time. Thus, it is very important to quantify the magnitudes of changes in panda habitat (an indicator for local biodiversity) caused by these factors. This chapter represents our attempt to examine the effects of demographic and socioeconomic variables on panda habitat in the Wolong Nature Reserve.

10.3 Long-Term Ecological Effects of Demographic and Socioeconomic Factors

We are interested in how changes in the demographic features (e.g., age structure, fertility) and socioeconomic conditions (electricity-related factors, particularly because of electricity's potential as a substitute for fuelwood) could affect panda habitat over a long time in a spatially explicit manner. Major questions of interest include (1) Which demographic and socioeconomic factors have significant (positive or negative) impacts on panda habitat? (2) How could economic factors, such as an electricity subsidy, conserve panda habitat? (3) How do spatiotemporal patterns of

[1]There were 49 (9 males and 40 females) people who migrated into and 67 (9 males and 58 females) people who migrated out of the reserve due to factors such as social networks established by seasonal workers.

panda habitat respond to changes in a combination of demographic and socioeconomic factors?

10.3.1 Design of Simulation Experiments

To answer the above questions, we designed a set of simulation experiments to understand how demographic and socioeconomic factors, when at play individually, would impact the two intermediate variables (population size and number of households), and our ultimate state variable (panda habitat), over space and time (Table 10.1). Through computer simulations, we tested a series of hypotheses (presented in Table 10.1) regarding the impacts of demographic and socioeconomic variables. First, for mortality and family planning factors, we examined the effects of mortality rates. We assumed that Wolong had been going through the same declining trend as the rest of China, reducing the mortality rates for all the six age groups by 50%. Second, we varied the fertility from 2.0 (average number of children allowed by the current policy in Wolong) to 1.0. This reduction is consistent with the fact that although Wolong currently has a higher fertility rate than cities in China, based on our field observations, many women in Wolong will tend to have fewer children in the future. Third, we examined the effects of birth interval by varying the length of birth interval (the time between the births of two consecutive siblings) from 3.5 to 8 years, corresponding to the "longer" part of the "fewer, longer, and later" family planning policy. Last, we examined the effects of marriage age by varying this age from 22 to 32 years old, corresponding to the "later" part of the policy.

To examine the effects of household formation and migration, we first evaluated the effects of "leaving parental home intention", the probability that a "parental-home dweller" (an adult child who remains in his/her parental household after marriage) would leave the parental household and set up a new household. We reduced the intention from 0.42 to 0.05, to encourage young adult children not to leave their parents' homes after marriage, which would probably result in larger household sizes and fewer households in the reserve. Second, we assessed the effects of education emigration – the migration of young people, aged 16–20 years, to college and other educational institutions outside the reserve (An et al. 2001). We used a variable "college rate" to indicate the proportion of children between 16 and 20 years old who could attend college. We varied the value of this variable from 1.92% to 50%, representing a policy alternative that could encourage more young people to move out of the reserve through approaches such as greater investment in education. This policy would be socially acceptable, due to the seniors' support of their children or grandchildren's outmigration to attend college (see Sect. 10.2). Last, we examined the effects of marriage migration, represented by a rise (from 0.28% to 50%) of "female marry-out rate", the ratio of the females between 22 and 30 years old who moved out of the reserve through marriage to all the females in this age group at a given year (An et al. 2005).

Table 10.1 Design of simulation experiments

Type of factors	Variable	Hypothetical impact[a]	Value of status quo	Individual experiment	Combined experiment	
				Changed value	Value for conservation scenario	Value for development scenario
Mortality and family planning factors	Mortality	+	Age dependent[b]	50% decrease		
	Fertility	–	2.0	1.0	1.0	4
	Birth interval	+	3.5	8.0	8.0	1.5
	Marriage age	+	22	32	32	22
Population movement factors	Leaving parental home intention	–	0.42	0.05	0.05	0.84
	College rate	+	1.92%	50%	50%	0.0%
	Female marry-out rate	+	0.28%	50%	50%	0.0%
Economic factors	Electricity Price	–	Location dependent	0.05 Yuan decline	0.05 Yuan decline	0.05 Yuan increase
	Outage levels	–	Location dependent	One level decrease	One level decrease	One level increase
	Voltage levels	+	Location dependent	One level increase	One level increase	One level decrease

[a]Hypothesized impact of a demographic or socioeconomic variable on the amount of panda habitat. A "+" sign means that a change in the value of a variable will change the amount of panda habitat in the same direction (e.g., a decrease in mortality will result in a reduction on panda habitat), whereas a "–" sign means a change in the opposite direction (a decrease in fertility will give rise to an increase in panda habitat)

[b] See An et al. (2005) for details

To assess long-term effects of economic factors, we reduced the cost of electricity by 0.05 Yuan kw^{-1} h^{-1} (US \$1 = 8.2 Yuan), reduced electricity outage by one level,[2] and increased voltage level by one level. These changes reflect the government's objectives of providing more high-quality electricity at a lower cost to substitute for the use of fuelwood. An "eco-hydropower plant" was recently constructed to achieve these objectives (M. Liu personal communication).

To understand the combined effects of various factors on population size, household numbers, and panda habitat, we designed a second set of simulation experiments with two opposing scenarios. The "Conservation Scenario" combined all the changes used in the above individual simulations that would help panda habitat conservation through decreases in human population, number of households, and fuelwood consumption. The "Development Scenario" set the values of all the related variables in the opposite direction (see Table 10.1), which would stimulate development of local economy and growth of local population and households. We chose a simulation period of 20 years for the economic factors, while we allowed demographic factors 30 years to take effect.

10.3.2 Model Description

To conduct the experiments outlined above, we used the Integrative Model for Simulating Household and Ecosystem Dynamics (IMSHED; An et al. 2005), which integrates various subsystems into a dynamic system that considers their interrelationships and the underlying mechanisms of various interactions from a systems perspective. IMSHED employs agent-based modeling (ABM) and geographic information systems (GIS). ABM can help predict or explain emergent higher-level phenomena by tracking the actions of multiple low-level "agents" that constitute or at least impact the system behavior observed at higher levels. Agents usually have some degree of self-awareness, intelligence, autonomous behavior, and knowledge of the environment and other agents as well; they can adjust their own actions in response to the changes in the environment and other agents (Lim et al. 2002). The model structure is illustrated in Fig. 10.4. IMSHED views individual persons and households as discrete agents and land pixels as objects. The layer of dashed households in the dashed box represents households at an earlier time, while the layer of solid ones represents households at a later time.

[2]Electrical outages had three levels: high, moderate, and low, representing more than 5, 2–4, and less than 2 outages per month, respectively. Voltage also had three levels, representing 220 V, 150–220 V, and fewer than 150 V (An et al. 2002). The default levels of outage and voltage for each household in the model were based on real data: values for a given household could be any of the three levels. If a specific household already has a low outage level, it would remain at that level regardless of the request of reducing outage level. Households with moderate or high levels of outage would have one level of reduction, resulting in low or moderate levels of outage, respectively.

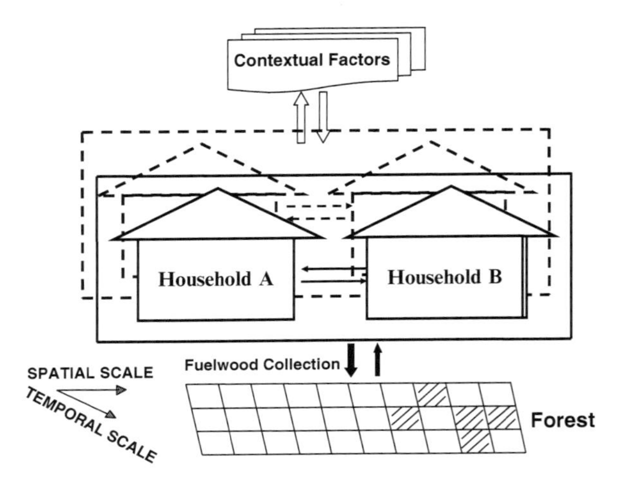

Fig. 10.4 Conceptual framework of the integrative model for simulating household and ecosystem dynamics (IMSHED; From An et al. 2005). Households A and B represent the households in Wolong, those within the *bold lines* refer to households at an earlier time and those within the *dashed lines* refer to households at a later time. The pixels at the *bottom* constitute the landscape of Wolong, where the *blank ones* are nonforest pixels and the *shaded ones* are forest pixels. Contextual factors include policy and geographical factors. The *arrows* stand for interactions between human population and the environment

The existing households (represented by households A and B in Fig. 10.4) come from the past and evolve into the future. They may increase or decrease in size, dissolve, or relocate. New households may be initiated as individual persons go through their life history. Household-level dynamics reflect individual-level events, such as birth, death, education emigration, and marriage migration.

A set of psychosocial factors determines or influences new household formation (An et al. 2003). Fuelwood demand is determined by a number of socioeconomic and demographic factors such as household size and cropland area (An et al. 2001, 2002). In the model, the forested landscape is divided into grid cells (pixels) to use the remotely sensed data and to facilitate simulations. In the process of fuelwood collection, a household evaluates the biophysical conditions (e.g., available forests and their topography), goes to an available pixel with the lowest perceived cost, and cuts trees for fuelwood. As a result, the forest pixel will be deforested and a nearby forest pixel will be chosen as the fuelwood collection site in the future. This process is part of the interaction between humans and the environment, as shown by the two arrows in Fig. 10.4. Initial conditions regarding tree species, growth rate, and total wood volume in each pixel are contained in the model. Contextual factors, including policies and geographical factors (e.g., elevation), also exert impacts on processes, such as household formation and fuelwood collection, and may ultimately impact panda habitat.

We used the data from 1996 to initialize all simulations. Although the total length of simulations ranged from 20 to 30 years depending on the objectives, the simulation time step was always 1 year. The model contains many stochastic processes, e.g., whether a person of a certain age group would survive a particular year depends on the number generated by the random number generator: if it is less than the associated yearly mortality rate, the person dies; otherwise he/she survives and moves to the next year. We ran a simulation 30 times (or replicates) to capture the variations among different replicates. We tested for the differences among various simulation results using two-sample t tests at the 0.05 significance level.

10.3.3 Simulation Results

Mortality and family planning factors affected population size, the number of households, and the area of panda habitat differently over a period of 30 years (Table 10.2). Compared to the baseline, population size significantly increased with a reduction in mortality (see Table 10.1 for their values in different simulations; the same for other variables hereafter) and decreased with a decline in fertility, a rise in birth interval, and an increase in marriage age (Fig. 10.5a). The number of households increased with the reduction in mortality, decreased with the decline in fertility and increase in marriage age, and remained nearly unchanged with the rise in birth interval ($p = 0.19$; also see Fig. 10.5a). With regard to panda habitat, the influences varied. All changes except the reduction in mortality ($p = 0.39$) significantly increased the amount of panda habitat (Fig. 10.6a).

Population movement (including migration and leaving parental home after marriage) factors affected population size, number of households, and panda habitat over 30 years significantly, with one exception (see Table 10.2). As expected,

Table 10.2 T-test ($\alpha = 0.05$) results in relation to the baseline situation[a]

Type of Factors	Variable	t-Statistic value (p value)		
		Population size	Number of households	Panda habitat
Mortality and family planning factors	Mortality	−34.32[b] (0.00)	−14.45 (0.00)	0.86 (0.39)
	Fertility	65.85 (0.00)	6.72 (0.00)	−2.52 (0.01)
	Birth interval	17.63 (0.00)	1.34 (0.19)	−2.04 (0.05)
	Marriage age	52.50 (0.00)	24.48 (0.00)	−13.70 (0.00)
Population movement factors	Leaving parental home intention	1.80 (0.08)	89.79 (0.00)	−22.81 (0.00)
	College rate	296.89 (0.00)	106.22 (0.00)	−25.73 (0.04)
	Female marry-out rate	95.76 (0.00)	38.48 (0.00)	−10.99 (0.09)
Economic factors	Electricity price	−1.31 (0.20)	−0.82 (0.43)	−46.70 (0.00)
	Voltage levels	0.79 (0.43)	−1.33 (0.194)	0.74 (0.46)
	Outage levels	0.13 (0.90)	0.40 (0.69)	−28.09 (0.00)

[a]Simulation lengths were 30 years for mortality and family planning as well as population movement factors, and 20 years for economic factors

[b]The "–" sign represents a reduction in value in the scenario compared to the associated value in the status quo, while no sign refers to an increase

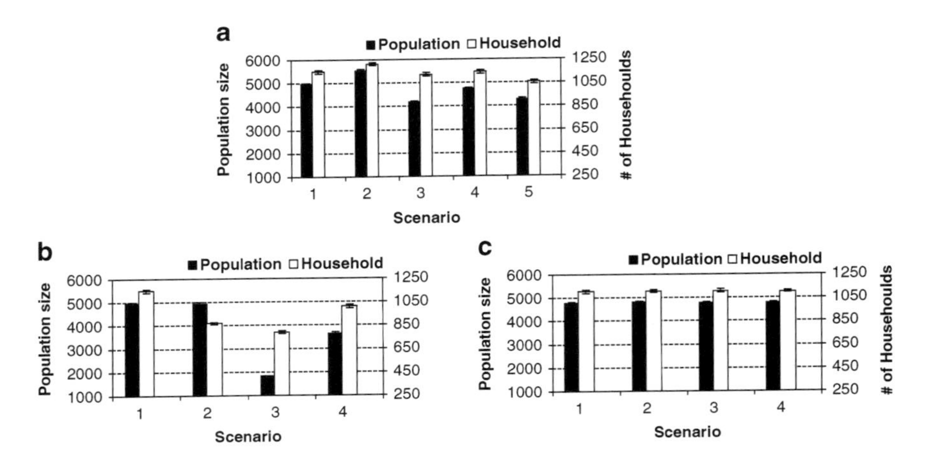

Fig. 10.5 Population size and the number of households in year 2026 (2016 for Economic factors) in response to changes in demographic and economic factors. (**a**) Family planning factors: scenario 1 for status quo, 2 for 50% reduction in mortality, 3 for fertility from 2.0 to 1.5, 4 for birth interval from 3.5 to 5.5, and 5 for marriage age from 22 to 28. (**b**) Population movement factors: scenario 1 for status quo, 2 for the intention of leaving-home from 0.42 to 0.21, 3 for college rate from 1.92% to 5%, and 4 for female marry-out rate from 0.28% to 20%. (**c**) Economic factors: scenario 1 for status quo, 2 for price reduction of 0.05 Yuan, 3 for one-level voltage increase, and 4 for one- level outage decrease. An *error bar* indicates one standard error

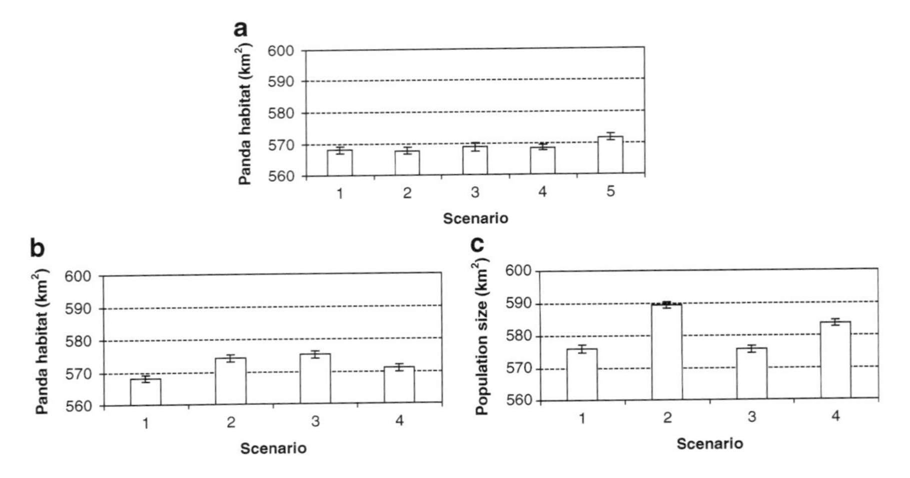

Fig. 10.6 Amount of panda habitat in year 2026 (2016 for Economic factors) in response to demographic and socioeconomic factors. (**a**) Family planning factors, (**b**) Population movement factors, and (**c**) Economic factors. The scenario definitions are identical to those in Fig. 10.5. An *error bar* indicates one standard error

the changes in the values of three population movement factors (a decrease in leaving parental home intention, an increase in rate of college attendance, and an increase in female marry-out rate) significantly reduced the number of households (see Fig. 10.5b). Their influence on population size varied, however. Leaving parental home had no statistically significant impact on population size ($p = 0.08$), while an increase in rate of college attendance and female marry-out rate significantly reduced population size ($p < 0.01$). The amount of panda habitat increased significantly (except for female marry-out rate with $p = 0.09$) as a result of increases in the values of all three factors (Fig. 10.6b).

The economic factors considered in our model had varying effects over a period of 20 years (see Table 10.2). The three scenarios (a decrease in electricity price, an increase in voltage level, and a decrease in the outage level) (see Fig. 10.5c) had insignificant influences on population size. Similarly, their impact on the number of households was insignificant. However, changes in the value of electricity price and outage level increased panda habitat significantly, while a change in the voltage level did not change the panda habitat significantly ($p = 0.46$; also see Fig. 10.6c).

A comparison between the Conservation Scenario and Development Scenario showed that substantial gaps existed between their projected impact on population size, number of households, and panda habitat area, which were significant and widened over time (Fig. 10.7a–c). At the end of 2026, there would be 5,300 fewer people, 1,000 fewer households, and 53 km^2 more panda habitat under the

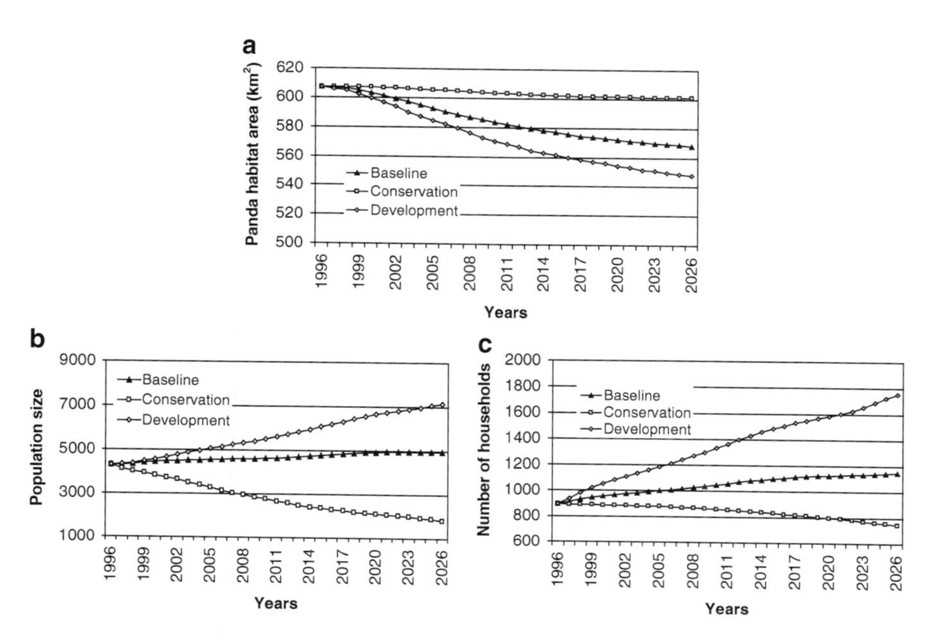

Fig. 10.7 (**a**) Panda habitat, (**b**) population size, and (**c**) the number of households under the status quo, conservation scenario, and development scenario over 30 years (1996–2026)

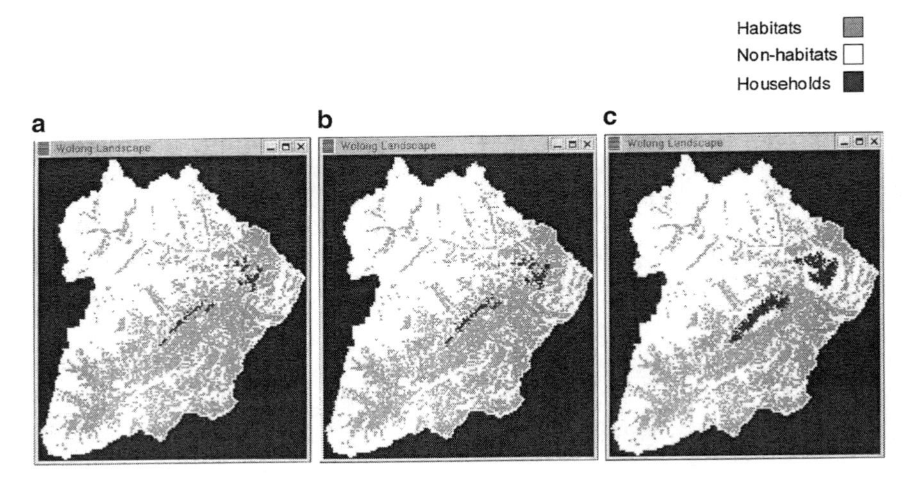

Fig. 10.8 Spatial distributions of panda habitat and households in 1996 and 2026 (desirable scenario, undesirable scenario) (An et al. 2005)

Conservation scenario than under the Development Scenario. When the spatial distributions of panda habitats and households were considered in the simulations (Fig. 10.8a–c), the impact caused by the demographic and socioeconomic factors became more apparent. For instance, much more panda habitat would be saved under the Conservation Scenario (Fig. 10.8b) compared to the Development Scenario (Fig. 10.8c), and the saved areas were located mainly near the households.

10.4 Discussion

According to our simulation results, mortality and family planning factors had a significant impact on population size, but a significant or insignificant impact on household dynamics and panda habitat. A change in mortality may take time to be translated into changes in population size and the number of households and ultimately into changes in panda habitat. The decline in fertility, the extension of birth interval between consecutive children, and delay in marriage age could reduce the number of new births, prolong the time between additional babies, and delay the birth of first babies (also increase the time between two generations), ultimately reducing demand for fuelwood, which may explain their significant effects in saving panda habitat. Although the magnitude (approximately 13.7 km^2 less habitat caused by an increase in marriage age) may be insubstantial compared to the total habitat area of 607.2 km^2 in 1996, it would make a greater difference when habitat distribution is considered. This study does not consider spatial factors, such as habitat fragmentation and the home range of pandas (2 km^2, according to Schaller et al. 1985). In our model, fragmented habitat (smaller than 2 km^2) has not been taken out. Thus, if a habitat of 100 km^2, for instance, were divided into small fragments of less than 2 km^2 each, the real loss would be 100 km^2 rather than zero.

Factors influencing population movement affected nearly all three response variables: population size, household numbers, and panda habitat. There were two exceptions: leaving-home intention and female marry-out rate showed no statistically significant effect on population size and panda habitat at the 0.05 significance level. Leaving-home intention largely deals with how likely it would be for a newly married couple to establish their own household, and it is not directly linked to population size. The female marry-out rate may need more time (i.e., longer than 30 years) to affect population size, household numbers, and, ultimately, panda habitat.

Economic factors had a significant impact on panda habitat because they encourage local residents to reduce their consumption of fuelwood by using more electricity. The Conservation Scenario and Development Scenario shed light on how demographic factors (especially those linked to population structure) and socioeconomic factors may influence panda habitat over time, illustrating substantial temporal and spatial differences in response to two opposite combinations of variables.

The results from our simulation study have important implications for the development of feasible and effective conservation policies. For example, promoting outmigration of young people through college education is not only socially desirable (Liu et al. 1999b), but also ecologically effective. Providing cheaper, more reliable, and higher-quality electricity for local residents could help them switch from fuelwood consumption to electricity use. A lesser dependence on fuelwood could help protect and restore panda habitat.

Future research should be directed towards the following aspects. First, it is necessary to collect more data (both cross-sectional and longitudinal) at the household level concerning demographic features, household economy (income and expenditures, material input/output), new household establishment, migration, and locations of both new households and fuelwood collection sites. These data could allow us to test hypotheses more rigorously in terms of plausible causal relationships (e.g., females' lower education could cause higher female out-migration through marriage). Such studies would not only be important to social scientists but could also be used to explain the dynamics of local biodiversity (represented by panda habitat in our study).

The impact of mortality and family planning factors on panda habitat may not be apparent for many years, as exemplified by the effects of an increase in marriage age on panda habitat. Thus, long-term studies of the factors presented in this chapter are essential. Spatial ABM (usually coupled with GIS) provides researchers with a useful tool to capture and integrate various detailed data – rather than just the averages – into a systems framework and overcome the shortcomings of traditional equation-based models. We conclude that this powerful experimental tool can promote a more complete understanding of long-term biodiversity dynamics across human-influenced landscapes.

Acknowledgments We thank Wolong Nature Reserve, especially Director Hemin Zhang for logistic assistance in field work and Jinyan Huang and Shiqiang Zhou for their assistance in data acquisition. We appreciate the valuable comments and edits from Richard Cincotta, Larry Gorenflo, and three anonymous reviewers. For financial support, we are indebted to the US National Science Foundation (NSF), US National Institutes of Health (NIH), American Association for Advancement of Sciences, John D. and Catherine T. MacArthur Foundation, and National Natural Science Foundation of China.

References

An L, Liu J, Ouyang Z, Linderman MA, Zhou S, Zhang H (2001) Simulating demographic and socioeconomic processes on household level and implications on giant panda habitat. Ecol Model 140:31–49

An L, Lupi F, Liu J, Linderman MA, Huang J (2002) Modeling the choice to switch from fuelwood to electricity: implications for giant panda habitat conservation. Ecol Econ 42(3):445–457

An L, Mertig A, Liu J (2003) Adolescents' leaving parental home: psychosocial correlates and implications for biodiversity conservation. Popul Environ 24(5):415–444

An L, Linderman MA, Shortridge A, Qi J, Liu J (2005) Exploring complexity in a human-environment system: an agent-based spatial model for multidisciplinary and multi-scale integration. Ann Assoc Am Geogr 95(1):54–79

Brashares JS, Arcese P, Sam MK (2001) Human demography and reserve size predict wildlife extinction in West Africa. Proc R Soc Lond Ser B Biol Sci 268(1484):2473–2478

China's Ministry of Forestry and World Wildlife Fund (1989) Conservation and management plan for Giant Pandas and their habitat. China's Ministry of Forestry and World Wildlife Fund, Beijing

Cincotta RP, Wisnewski J, Engelman R (2000) Human population in the biodiversity hotspots. Nature 404:990–992

Cooney RS, Shi J (1999) Household extension of the elderly in China, 1987. Popul Res Policy Rev 18(5):451–471

Fang RK (1993) The geographical inequalities of mortality in China. Social Sci Med 36(10): 1319–1323

Feng G, Hao L (1992) Summary of 28 regional birth planning regulations in China. Popul Res (Renkou Yanjiu; in Chinese) 4:28–33

Forester DJ, Machlis GE (1996) Modeling human factors that affect the loss of biodiversity. Conserv Biol 10(4):1253–1263

Hussain A (2002) Demographic transition in China and its implication. World Dev 30(10): 1823–1834

Lim K, Deadman PJ, Moran E, Brondizio E, McCracken S (2002) Agent-based simulations of household decision-making and land use change near Altamira, Brazil. In: Gimblett HR (ed) Integrating geographic information systems and agent-based techniques for simulating social and ecological processes. Oxford University Press, New York, pp 277–308

Liu J, Taylor WW (eds) (2002) Integrating landscape ecology into natural resource management. Cambridge University Press, Cambridge, UK

Liu J, Ouyang Z, Tan Y, Yang J, Zhang H (1999a) Changes in human population structure: implications for biodiversity. Popul Environ 21:46–58

Liu J, Ouyang Z, Taylor WW, Groop R, Zhang H (1999b) A framework for evaluating the effects of human factors ob wildlife habitat: the case of giant pandas. Conserv Biol 13:1360–1370

Liu J, Linderman M, Ouyang Z, An L, Yang J, Zhang H (2001) Ecological degradation in protected areas: the case of Wolong Nature Reserve for giant pandas. Science 292:98–101

Liu J, Daily GC, Ehrlich PR, Luck GW (2003) Effects of household dynamics on resource consumption and biodiversity. Nature 421(6922):530–533

Ma Z (1999) Temporary migration and regional development in China. Environ Plann A 31:783–802

McKee JK, Sciulli PW, Fooce CD, Waite TA (2004) Forecasting global biodiversity threats associated with human population growth. Biol Conserv 115(1):161–164

Mittermeier RA, Gil PR, Hoffman M, Pilgrim J, Brooks T, Mittermeier CG, Lamoreux J, da Fonseca GAB (2005) Hotspots revisited. Conservation International, Washington, DC

Myers N, Mittermeier RA, Mittermeier CG, da Fonseca GAB, Kent J (2000) Biodiversity hotspots for conservation priorities. Nature 403:853–858

Schaller GB, Hu J, Pan W, Zhu J (1985) The giant pandas of Wolong. University of Chicago Press, Chicago and London

Tan Y, Ouyang Z, Zhang H (1995) Spatial characteristics of biodiversity in Wolong Nature Reserve. China's Biodiversity Reserve 3:19–24 (in Chinese)

Veech JA (2003) Incorporating socioeconomic factors into the analysis of biodiversity hotspots. Appl Geogr 23(1):73–88

Wolong Nature Reserve (1997) Wolong Agricultural Operation and Contracting Situations (unpublished, in Chinese)

Wolong Nature Reserve (2000) Wolong 2000 Nationwide Population Census Form (unpublished, in Chinese)

Chapter 11
Exploring the Association Between People and Deforestation in Madagascar

L.J. Gorenflo, Catherine Corson, Kenneth M. Chomitz, Grady Harper, Miroslav Honzák, and Berk Özler

11.1 Introduction

Few countries in the world can match the contribution to global biodiversity made by the small island nation of Madagascar (Myers et al. 2000). Due to its geologic separation from the prehistoric super-continent of Gondwanaland between 160 and 70 million years ago (Smith et al. 1994), much of the flora and fauna of Madagascar evolved in geographic isolation from other parts of the world. The result of this isolated evolution has been a biological inventory characterized by particularly high endemism – plants and animals found nowhere else – including 33 species of lemurs, more than 10,800 species of plants, 215 species of amphibians, in excess of 300 species of reptiles, and more than 100 species of birds (Langrand and Wilmé 1997; Mittermeier et al. 1994, 2004; Morris and Hawkins 1998; Schatz 2000; Vallan 2000). In addition, Madagascar is home to particularly high diversity of certain biological resources, notably reptiles and amphibians (Mittermeier et al. 2004; Vallan 2000). As much as 90% of Madagascar's endemic biota live in forest ecosystems (Dufils 2003), making remnants of the forests that once covered much of the island of great importance to conservation.

The remaining biodiversity in Madagascar is under considerable pressure. A rapidly growing population, considerable internal migration, reliance on crop production that often requires frequent agricultural expansion to compensate for poor soils, conflicting customary and state-defined land tenure, increasing reliance on forest resources to meet local and urban demands for timber and fuel, and rapid expansion of mining for minerals and gemstones all contribute to the alteration or destruction of natural habitat (Durbin et al. 2003; Erdmann 2003; Fischer et al. 2002; Kistler and Spack 2003; Population Reference Bureau 2004; Vincelette et al. 2003; World Bank 2005). Numerous extinctions linked to human exploitation or human-induced habitat destruction over the past several centuries attest to the potential consequences of human impact (Dewar 1984, 1997, 2003; Goodman et al. 2003), including the disappearance of at least 15 species of lemurs and other well-known animals, such as the giant elephant bird (Mittermeier et al. 1999).

R.P. Cincotta and L.J. Gorenflo (eds.), *Human Population: Its Influences on Biological Diversity*, Ecological Studies 214,
DOI 10.1007/978-3-642-16707-2_11,

Successful conservation on Madagascar requires an increased understanding of human impacts on remaining biodiversity – a first crucial step in *addressing* the causes of biodiversity loss through shifts in policy or development strategies. To advance such understanding, this chapter examines factors potentially associated with deforestation in this island nation during the 1990s, with the continued reduction of Madagascar's remaining forests generally considered the most important cause of biodiversity loss in the country. Although forest loss in many ways reflects the demands of human population, in the following study we find that there is no simple relationship between deforestation and human demographics. In an attempt to provide the basis for introducing development policies that enable the conservation of biodiversity while meeting essential human needs, we consider several other economic, geographic, and institutional factors in addition to population. Results of statistical analyses indicate that the relationship between deforestation and both physical and socioeconomic conditions in Madagascar is complex, often varying by region. Nevertheless, our study identifies variables that are both associated with deforestation and sensitive to policy decisions, introducing the potential to stem forest loss through strategic development.

11.2 Madagascar: Selected Characteristics of the Physical and Human Setting

Lying about 400 km off the east coast of Africa, Madagascar is the world's fourth largest island, covering about 587,000 km^2 (Jolly et al. 1984; Hance 1975; Preston-Mafham 1991). Physical geography and climate combine to produce an extremely diverse natural environment. Much of the main island of Madagascar consists of an upland plateau dissected by systems of hills and valleys formed by millennia of geologic forces and erosion. To the east, the central plateau descends abruptly to a narrow coastal shelf, while the descent to the sea is much more gradual to the north, west, and south. Red lateritic clay overlies much of the bedrock on Madagascar's central plateau, with soils in the western third of the island increasingly derived from weathered sedimentary formations. Madagascar's climate generally is tropical, though both temperature and rainfall patterns vary greatly across the island (Donque 1972). Rainfall is heavy on the east coast and in the northwest, where it falls year-round, and much less in the west and (especially) southwest where seasonal precipitation occurs. The geographic pattern of vegetation reflects differences in topography and climate, with lowland and montane rainforest occurring in the east, Sambirano subhumid forest in the northwest, dry deciduous forest and savanna in the interior and west, and semiarid thicket in the southwest (Gautier and Goodman 2003) (Fig. 11.1). This regional diversity is reflected in regional heterogeneity in human–environment relations and, in particular, in causes of deforestation.

The population of Madagascar is growing rapidly, at a rate of about 2.7% annually, due primarily to persisting high fertility (total fertility rate of 5.2) (United

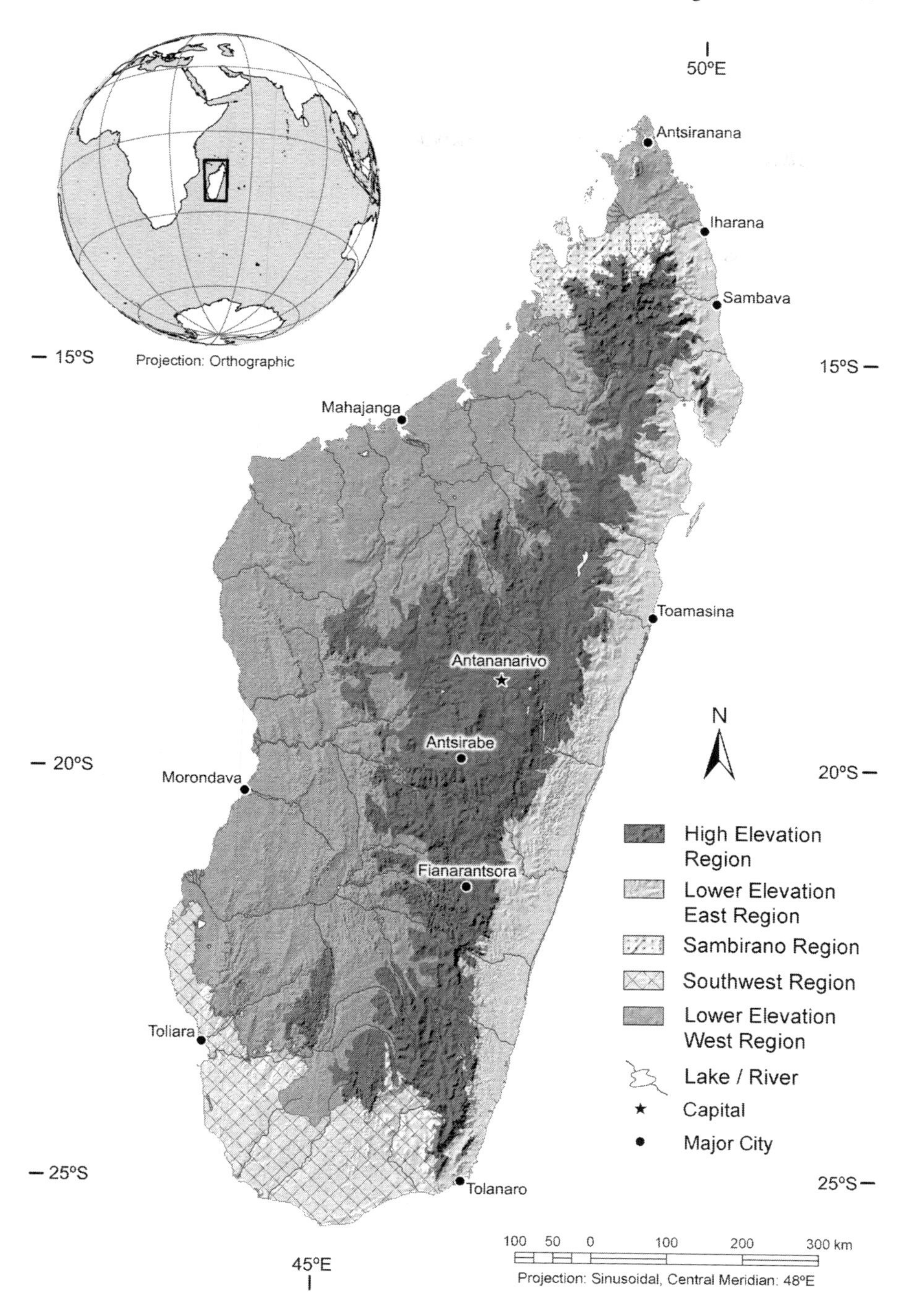

Fig. 11.1 Madagascar: selected physical and human geographic characteristics

Nations Population Division 2005). As this manuscript was completed, the most recent census was more than a decade old, that 1993 effort recording 12.3 million total inhabitants (Bureau Central du Recensement 1993). The United Nations estimated the 2005 population of Madagascar as 18.6 million (United Nations Population Division 2005). Only about 20% of the current population of Madagascar lives in urban settings, the remainder residing in smaller settlements dispersed across the rural countryside (Kistler and Spack 2003) (Fig. 11.2). Population projections prepared by the United Nations for 2050 range between 37 and 51 million, depending on assumptions about mortality and fertility (United Nations Population Division 2005).

Madagascar is among the world's poorest countries, with a gross national income per capita at about US$290 (World Bank 2005). The headcount poverty rate is greater than 70%, with poverty rates of 75% or more in the rural areas of most provinces (based on a poverty threshold of 354,000 Malagasy francs per capita household expenditure per month) (Mistiaen et al. 2002). The majority of Madagascar's residents participate in a rural mixed economy involving small-scale subsistence agriculture combined with some cash crop production, forest resource commerce, or cattle herding. Larger scale agriculture involving commercial rice, maize, coffee, sisal, and sugar production, along with mining and timber extraction, also occur (Durbin et al. 2003; World Bank 2003).

Small-scale, shifting slash-and-burn agriculture, called *tavy* in the eastern rainforest (though technically this refers only to upland rice agriculture – see Bertrand and Sourdat 1998) and *hatsake* in the southwest (usually involving maize – see Casse et al. 2004), is common in Madagascar. Unfortunately, fertility of the poor lateritic and calcareous soils often is quickly exhausted, requiring subsistence agriculturalists to cultivate new areas every couple of years. With adequate fallow periods, in the past shifting agriculture in Madagascar was sustainable as farmers allowed previously used fields to regenerate their nutrients and farmed them again (Rakotomanana 1989). However, increased population in particular areas has led to shortened fallow periods, causing soil exhaustion and requiring that farmers expand increasingly into natural habitat for crop production (Brand and Zurbuchen 1997; Jolly 1989; Laub-Fischer et al. 1997).

Commercial crop production is important in Madagascar, providing about half of the foreign exchange earnings in the country (Kistler and Spack 2003). But growing cash crops faces some of the same problems as subsistence agriculture, notably soil exhaustion in the absence of nutrient replenishment through the application of fertilizer. Moreover, commercial production of some crops often involves plowing, which leads to considerable erosion when fields occur on hillsides. Commercial agriculture is expanding, in part due to growing demand for increasing amounts of cash brought about by economic conditions such as inflation (see Economist Intelligence Unit 1998), as well as growing demand in towns and abroad for agricultural products (Durbin et al. 2003; Ministère du Développement Rural et de la Réforme Agraire 1995).

Another widespread economic activity in rural Madagascar is the direct exploitation of forest resources. In particular, forests represent important sources of fuel (in the form of wood and charcoal) and construction material. Although some

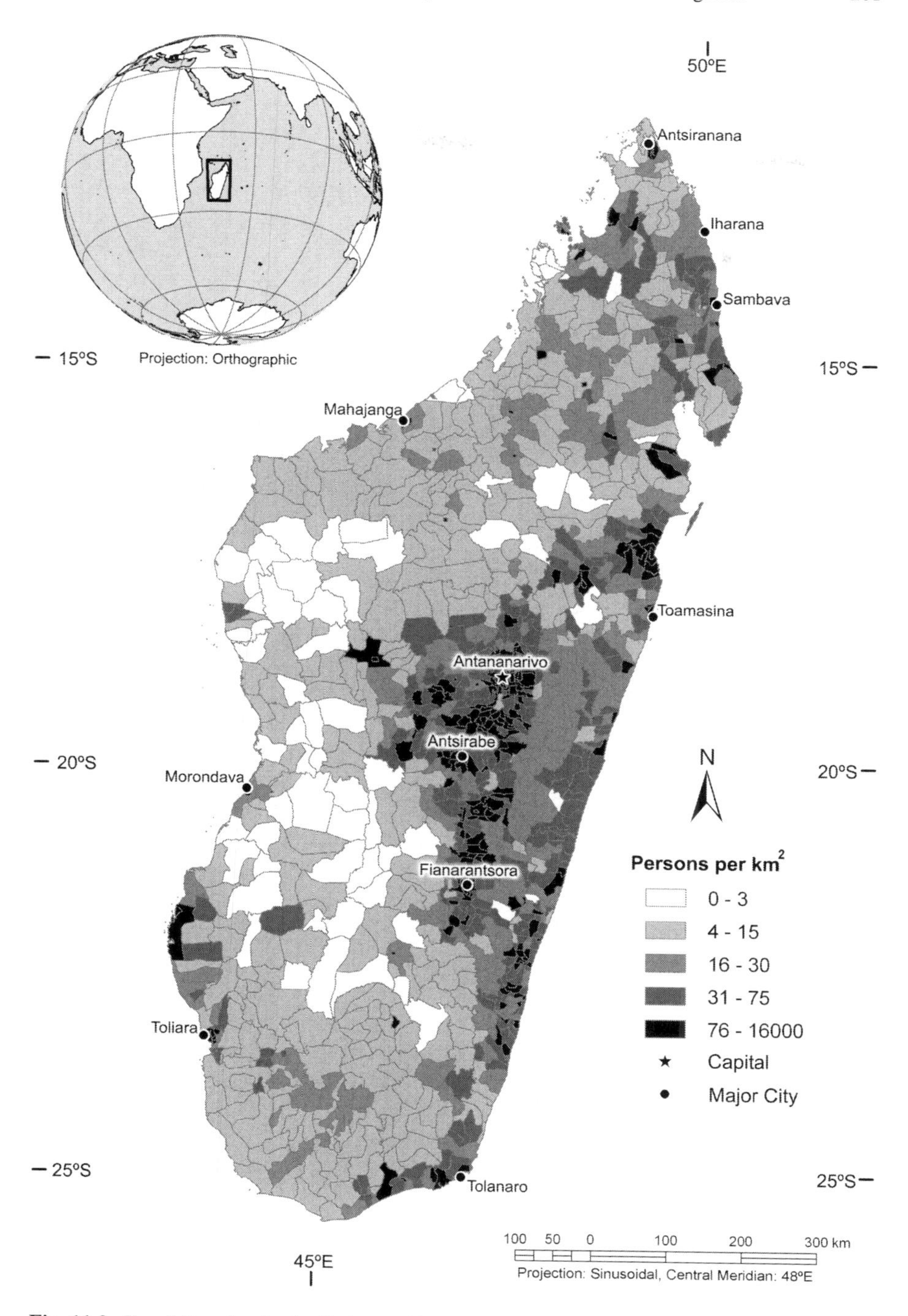

Fig. 11.2 Population density by firaisana, 1993

eucalyptus and other types of tree plantations exist to supply such resources, in certain cases large population centers rely on natural forests for more than 90% of their wood requirements (Abel-Ratovo et al. 2000). Indeed, meeting the demands of urban centers for fuel and construction material has emerged as a major cause of deforestation in Madagascar (Durbin et al. 2003). In the most severe cases of forest resource extraction, entire stands of trees are felled. However, in many instances of fuel harvesting the main impact is forest degradation rather than destruction, as selected parts of trees are removed (Casse et al. 2004).

Herding occurs in much of rural Madagascar, primarily of cattle but also of sheep and goats (the latter infrequent) (Kistler and Spack 2003). Herding is important both economically, as a means of supplying key resources such as protein, and symbolically, as a representation of status through capital accumulation (Kistler and Spack 2003; Ratsirarson 2003). Much herding occurs in the central highlands of Madagascar, on grasslands managed by setting fires to generate regrowth of plants fed upon by livestock (Kull 2003). But herding also occurs in other parts of the country and in other environmental settings – including forests, where free ranging animals eat grass in open areas and feed on tree seedlings (Ratsirarson 2003).

Mining in Madagascar involves the extraction of several natural resources, such as gems, graphite, marble, sand, and titanium (Goodman and Benstead 2003). The mining sector in Madagascar comprises two subsectors: the large-scale mining sector, which focuses primarily on various industrial ores, and the small-scale informal, unregulated gemstone mining sector, pursued by individuals or small groups in search of gems (Cardiff and Andriamanalina 2007; Cardiff and Befourouack 2003; Duffy 2005; Sarrasin 2006; Walsh 2004). Small-scale mining is widespread and increasing. Large-scale mining is less widespread, though it also is on the rise. Mining has direct and indirect adverse impacts on the natural environment of Madagascar – converting habitat at the mine sites, generating contamination through runoff from extracting and processing resources, and attracting in-migrants who in turn place demands on local resources.

The physical and human geographies of Madagascar present limited economic alternatives, many of which have led to the removal of natural forest. Although previous research has identified patterns of forest loss, to date there has not been a systematic, countrywide study of the conditions under which deforestation occurs. In the interest of identifying conditions associated with recent patterns of forest removal that development policy might influence, in the pages that follow we examine deforestation patterns during the 1990s.

11.3 Recent Patterns of Deforestation in Madagascar

Deforestation has long been considered a problem in Madagascar. As early as the 1800s, Malagasy kings recognized the detrimental effects of deforestation and burning in their country (Wright 1997). French colonial authorities similarly noted the environmental damage associated with deforestation during the late nineteenth and early twentieth centuries. Ironically, although they established regulations against shifting cultivation, they also granted timber concessions and expropriated nearly one million ha of land from peasants for the production of export crops, greatly accelerating forest loss (Jarosz 1993). Uncertainty surrounding key pieces of information, including the degree to which the highlands were once forested (Gade 1996; Klein 2002; McConnell 2002), affects our understanding of forest loss. Varying estimates of deforestation in Madagascar beginning in the mid-twentieth century, although not in total agreement, indicate the enormous magnitude of forest loss over the past several decades (see Dufils 2003; Myers 1986; Harper et al. 2007).

Deforestation reduces biodiversity not only through habitat loss, but also through the fragmentation of forested areas. The reduction of contiguous forest and patch size can have a negative impact on certain birds (Langrand and Wilmé 1997), insectivorous mammals (Goodman and Rakotondravony 2000), and amphibians (Vallan 2000) while increasing predator access (Smith et al. 1997). Unfortunately, the consequences of forest loss often are long-term. The potential for regenerating forest habitat in much of Madagascar is low due to the massive erosion that often accompanies deforestation – with research reporting annual watershed erosion rates as high as 250 tons per ha (Helfert and Wood 1986).

Madagascar's tremendous biological endemism, and the threat it faces from ongoing deforestation, recently led researchers at Conservation International to map forest cover and patterns of forest loss between 1990 and 2000 for the entire island (Harper n.d.; Harper et al. 2007). Using Landsat 5 and 7 satellite imagery from the 1990 and 2000 periods,[1] analysts identified and mapped forest cover and changes in humid and montane forest in the east and north, dry forest in the west, spiny forest in the south, and mangroves. That study designated changes in a given satellite image pixel from forest in 1990 to nonforest in 2000 as deforestation, providing the information needed to map forest loss during the decade and define the rate of deforestation (Fig. 11.3).

The results of analyzing forest cover indicated that more than 106,000 km^2 of Madagascar were forested in 1990, about 18.1% of the terrestrial area (Table 11.1; see Fig. 11.3). Of the total forest cover that year, humid forest accounted for about half (49.4%), with dry and spiny forest each composing about one-fourth of the 1990 total. By 2000, the total forested area had declined by 8.6% to less than 90,000 km^2. Although forest still accounted for more than 15.3% of land cover, nearly 8,100 km^2 of forest had been lost in a single decade (Harper et al. 2007).

[1]The analysis of satellite imagery sought to identify deforestation between 1990 and 2000. However, as often is the case with islands clouds on parts of images occasionally obscured the ground. As a result, for some locations analysts had to analyze imagery dating slightly earlier or later than the 1990 and 2000 target dates in order to fill in these clouded areas.

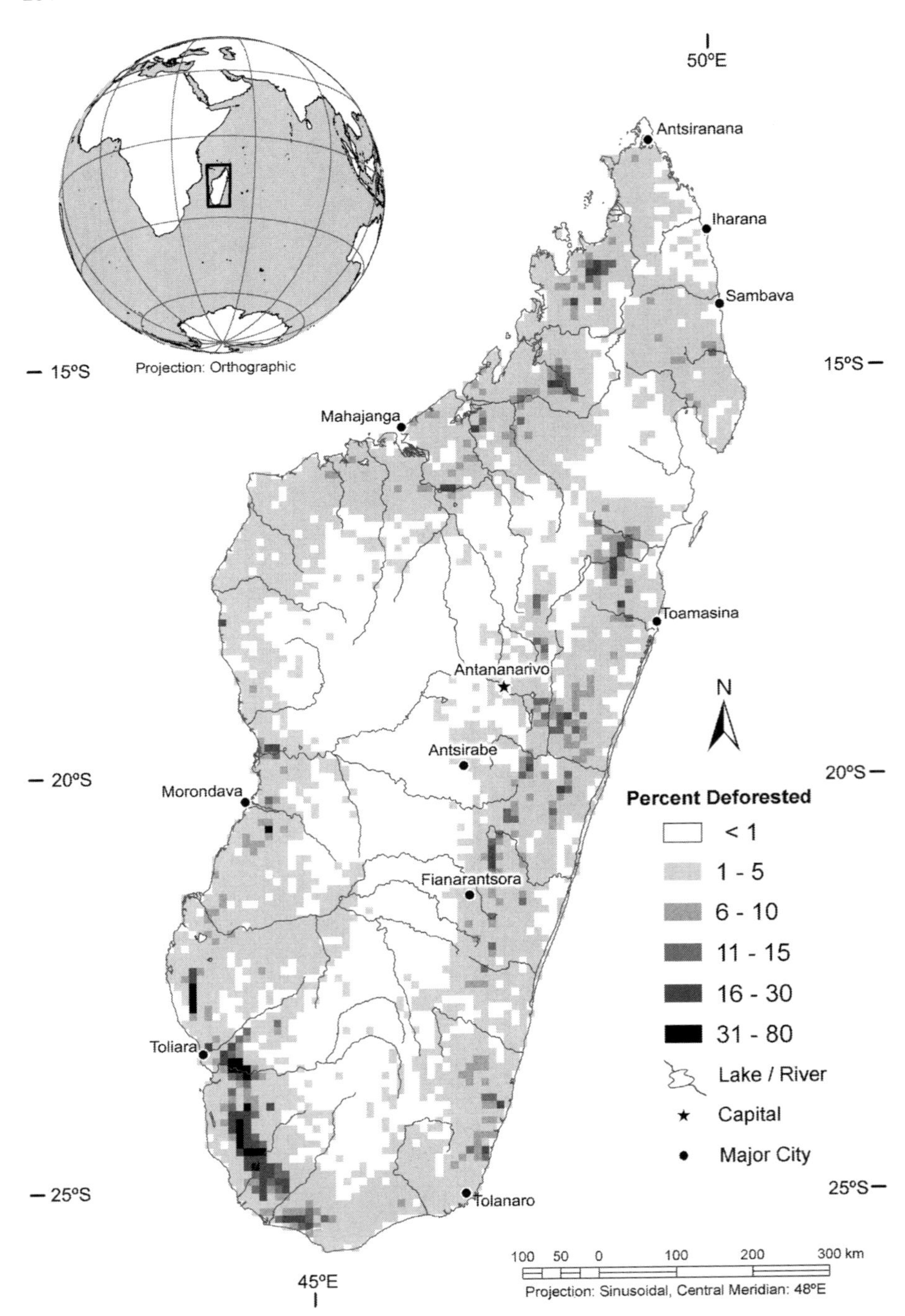

Fig. 11.3 Deforestation, 1990–2000, shown as percent of pixels in 10 × 10 km cells forested in 1990 that lacked forest in 2000. Note that the coarse resolution of this map is for illustrative purposes only; the resolution of deforestation data used in the analysis was 28.5 m

Table 11.1 Deforestation in Madagascar, 1990–2000

Forest type	1990 Known forest (km^2)[a]	2000 Known forest (km^2)[a]	Deforestation 1990–2000 (km^2)[b]	Percent deforested 1990–2000[b]
Humid	52,343	41,668	3,220	7.8
Dry	27,118	24,570	1,982	7.5
Spiny	24,200	21,322	2,817	11.7
Mangrove	2,396	2,261	55	2.4
Total	106,057	89,821	8,074	8.6

[a]"Known forest" refers to portions in a satellite image that can be defined as forest *plus* portions that, although covered by clouds in a particular image, are revealed to be forest in an image of a later date

[b]Deforestation amounts and deforestation rates presented consider only areas visible in 1990 and 2000 imagery – that is, cloud-free and shown to be forest in 1990 and without forest in 2000. This provides more accurate estimates of deforestation because it avoids errors where cloud-covered forest in 1990 was cleared prior to 2000 and, thus, because it does not appear as forested in 2000 would not be categorized as forested in 1990

11.4 Analyzing Patterns of Deforestation in Madagascar

11.4.1 Previous Evaluations of Deforestation in Madagascar

Although people have been concerned with deforestation in Madagascar for centuries, developing an understanding of this process in such a complex environmental and socioeconomic setting has remained elusive. Before we discuss our analysis of recent patterns of deforestation in Madagascar, let us consider briefly the results of selected earlier studies and the insights they provide, both in terms of identifying key variables and in proposing the relationship of these variables with forest loss.

Many researchers argue that agriculture is the primary cause of deforestation in modern Madagascar. Often the blame is placed on the shifting agriculture of subsistence farmers (Green and Sussman 1990; Jolly 1989; Messerli 2002; Rakotonindrina 1988), primarily in the humid zone of the east (Erdmann 2003) but also in drier parts of the island (Abel-Ratovo et al. 2000; Cabalzar 1990). However, research indicates that commercial agriculture of crops such as maize, butter beans, groundnuts, sisal, and tobacco also contributes to deforestation (see Durbin et al. 2003; Kistler and Spack 2003). The role of agriculture in deforestation often is complex, in that it can involve many factors. For instance, a recent statistical analysis of deforestation in southwestern Madagascar argued that expanding agriculture, primarily for export crop production, is the main *proximate* cause of deforestation in recent years for this part of the country (Casse et al. 2004). That study identified crop production integrated with other direct (wood collecting and herding) and indirect causes (migration, export prices, property rights, and governmental policies) – many also associated with agriculture – as playing a key role in forest loss.

Human population is another variable found to be associated with deforestation in certain parts of Madagascar. Focusing on the eastern rainforests, Sussman et al. (1994)

found a positive correlation between population and forest loss – with lessened deforestation of rainforest in areas with sparse population. Another evaluation of deforestation on the eastern escarpment also showed a positive relationship between population and forest loss, with a time series revealing dramatic deforestation coinciding with a large increase in human habitation (Messerli 2002). But research on the relationship between population and deforestation in many ways remains inconclusive. For instance, more rapid deforestation in parts of the eastern rainforests occurred for a time in more densely populated areas and then in less densely populated areas – the shift apparently due to decreasing accessibility of forest in densely populated areas as people harvested inventories over time (Green and Sussman 1990). In southwestern Madagascar, neither population nor migration adequately explained patterns of deforestation (Casse et al. 2004). Historically, some of the most rapid instances of deforestation occurred when population growth was low, even below replacement rate (Jarosz 1993). Upon closer examination, various aspects of human behavior likely complicate the relationship between human demographics and deforestation. As an example, increased attractiveness of some locality due to a new mine may increase population and deforestation – the latter possibly a consequence of mining operations as much as increased human presence. Ultimately, the association between human population and deforestation depends largely on the activities in which humans are involved – the *per capita* impact associated with certain behavior often much greater than that associated with another.

Access also has been proposed as an important factor in deforestation in Madagascar, with proximity to roads in eastern Madagascar stimulating higher deforestation [Rakotomamonjy 2000 (cited in Kistler and Spack 2003)]. The role of access is particularly important for commercial agriculture, as it provides the means of transporting products to markets (Kistler and Spack 2003). In contrast, Freudenberger (2003) argues that, in the case of the Fianarantsoa-East Coast Railroad, improved access shifted local populations away from deforestation in favor of geographically concentrated commercial tree–crop agriculture.

Various studies have proposed selected other socioeconomic factors as associated with deforestation in Madagascar, though these conclusions tend to be based more on qualitative observations and assessments rather than quantitative analytical evaluations. In their review of socioeconomic factors associated with biodiversity loss, Durbin et al. (2003) identify demand for forest products (primarily fuel wood, charcoal, and timber), migration of people from elsewhere in Madagascar, and commercial agriculture as major concerns. Insecure land tenure that enables the influx of migrants to some areas, often to practice commercial or subsistence agriculture or to earn wages for various forms of labor, also has been argued as associated with deforestation (Kistler and Spack 2003). Customary rules and laws, as well as modern laws, can restrict deforestation, but the degree of enforcement varies considerably across the country (Horning 2003).

Previous studies of deforestation in Madagascar provide no clear conclusions about the preconditions of forest loss in this island nation. Research tends to be geographically restricted to subsections of the country or to focus on single variables as opposed to multiple potential causes. What appears to be associated with deforestation in one locality often does not show a similar association elsewhere. To increase our shared understanding of forest loss, we explore the potential role of several variables in recent patterns of deforestation – focusing initially on the entire island of Madagascar and then on individual regions.

11.4.2 Methods and Data

To evaluate the relationship between deforestation and variables potentially associated with it, we conducted a statistical analysis using a multivariate probit model to examine deforestation between 1990 and 2000. We applied the model to 93,093 points arranged at 1-km intervals across Madagascar, focusing exclusively on localities that were forested in 1990 – that is, places where deforestation would have been possible over the ensuing decade. For each point examined, forest cover change (1990–2000) assumes the role of dependent variable and takes on the value *remained forested* or *was deforested* to enable evaluation of the statistical relationship between forest cover and several independent variables. In all, 8.4% of the points examined became deforested during the 1990s, the slight discrepancy with the 8.6% deforestation rate calculated from satellite imagery (see Harper et al. 2007) due mainly to differences in dealing with areas obscured by clouds in 1990 and forest in 2000. We considered several potential predictors of forest loss during the 1990s: (1) socioeconomic characteristics of population density and poverty; (2) access to forested areas, via roads and footpaths; (3) physical geographic factors, including elevation, slope, and soil fertility constraints; and (4) protected status, which to varying degrees (depending on the type of protected area and level of enforcement) constrain resource extraction and other types of human uses. In addition, by dividing the country into five agroclimatic zones – High Elevation (over 800 m above sea level), Southwest (primarily characterized by dry, spiny forest), Low Elevation East, Low Elevation West, and the Sambirano area of (originally) lowland rainforest in the northwest (see Fig. 11.1) – we examined how the association between certain variables and deforestation varies regionally. We conducted statistical analyses for forests across Madagascar, controlling for regional variation by including regional dummy variables, and then conducted separate analyses for each of the five regions.

The multivariate probit model used in our analyses takes the following form:

$D^* = \alpha + \beta_1$ population density + β_2 distance to nearest road + β_3 distance to nearest footpath + β_4 mean monthly household expenditure + β_5 household income inequality + β_6 topographic slope + β_7 soil fertility constraints + β_8 presence of protected area + β_9 elevation + Σy_j region$_j$ + ε

$$D = 1 \quad \text{if } D* > 0, D = 0 \quad \text{if } D* \leq 0,$$

where D is the observed dependent variable, signifying deforestation between 1990 and 2000, classified as 1 if deforestation occurred at a particular locality, 0 if it did not; α is a constant; D^* is a latent variable based on the observed value of D; β_i is a parameter corresponding to each independent variable i; y_j is a parameter associated with a dummy variable for region j; and ε is an error term.

We obtained data on deforestation from satellite imagery interpreted and analyzed by Conservation International for 1990 and 2000 (Harper n.d.; Harper et al. 2007). Values for the independent variables considered in the model came from the following sources:

- Population density: The 1993 census of population and housing, with data presented at the firaisana level – the smallest administrative units used to present data in the 1993 census (Bureau Central du Recensement 1993).
- Poverty: A poverty map for Madagascar, constructed by World Bank and Malagasy researchers, which estimates consumption-based welfare from the 1993 census of population and housing and a 1993 household survey, again presented at the firaisana level (see Mistiaen et al. 2002).
- Roads and footpaths: L'Institut Géographique et Hydrographique National de Madagascar, the national mapping center of Madagascar, who provided data on roads and paths at a scale of 1:500,000 based on remote sensing imagery from the early 1990s.
- Slope and elevation: L'Institut Géographique et Hydrographique National de Madagascar, the 100-m, vertically spaced contour data providing the basis for a digital elevation model from which we derived slopes using standard geographic information system techniques (Zeiler 1999).
- Soil fertility constraints: Global Agro-Ecological Zone assessment, based on evaluating data from the Food and Agriculture Organization soil map of the world in the context of various crop production scenarios (Fischer et al. 2002).
- Protected areas: Madagascar Association Nationale pour la Gestion des Aires Protégées – our focus limited to parks established before 1990 (that is, protected for the entire period for which we examined forest cover change) and including national parks, strict nature reserves, special reserves, and marine parks with a coastal terrestrial component.

All of the data used in this study have geographic coordinates, enabling their analysis with respect to spatial proximity to one another. However, the spatial resolution varied considerably among datasets: 28.5-m cells for deforestation, 100-m cells for elevation and slope, 9-km cells for soil fertility constraints, and units (firaisana) averaging 476 km^2 for population and poverty. Although it would be possible to aggregate all data to a common geographic unit – in this case the firaisana – such aggregation would sacrifice considerable geographic detail on deforestation patterns, access, elevation, and agricultural constraints. As an alternative, to preserve the locational detail present in much of the data, for all analyses with deforestation as the dependent variable, we used the set of 93,093 sample points. For independent socioeconomic variables available only in geographic units larger than 28.5 m, we used the mean values for the unit in which each sample point occurred; for remaining independent variables, we used the locational information available in the dataset (e.g., to calculate distance between a forested or deforested cell and the nearest road). The modeling process, which involved considering several variables concurrently, conceptually involves "... [stacking] the data layers (maps) of interest. A pin pierces the stack at each sample point, and the mapped information for the point – slope, distance to road, soil quality – is recorded and collated" (Chomitz and Gray 1996).

11.4.3 Results

Results of the multivariate probit analysis for all of Madagascar appear in Table 11.2. This table reports the impact on the probability of deforestation from a unit change in each independent variable – presented in the table as *marginal effects*. The association of variables with deforestation and the effect of each on the likelihood of forest loss vary considerably.

Although statistically significant (at $p = 0.001$), the actual impact of changes in population density is minimal. An increase in local population density of 10 persons per km^2 increases the absolute probability of deforestation rate by 0.15% points per decade. Because the deforestation rate in Madagascar between 1990 and 2000 for all sample points was 8.4%, such a change amounts to a relative

Table 11.2 Statistical results of probit modeling of deforestation for all of Madagascar by independent variable

Variable	Marginal effect[a, b]	Standard error
Distance to nearest footpath[c]	−0.00000381	0.0000003
Distance to nearest road[c]	−0.00000226	0.0000001
Elevation[d]	−0.0000119	−0.000004
Inequality[e]	−0.0396	0.02
Mean monthly per capita household expenditure[f]	−0.0000432	0.00001
Population density[g]	0.000146	0.00003
Protected area[h]	−0.0515	0.003
Slope[i]	−0.00142	0.0001
Soil fertility constraints[j]	0.000565	0.00006

[a]Marginal effect is the change in predicted probability of deforestation associated with a unit change in a particular independent variable, holding other variables at sample means; for the discrete variable *protected area*, the marginal effect is the difference in probability of deforestation associated with the presence vs. absence of a protected area

[b]All coefficients are significant at a 0.001 level, except for Inequality which is significant at a 0.05 level

[c]Meters

[d]Meters above sea level

[e]Gini coefficient, defined as the ratio between existing income distribution and theoretical income equality across a population. For a perfectly equal distribution, the Gini coefficient equals zero. For complete inequality (where only one person has any income), the Gini coefficient is one

[f]000s of Malagasy francs

[g]Persons per km^2

[h]Presence/absence

[i]Degrees

[j]Organized as 7 constraint categories, ranging from "no constraints" (with an index of 0) to "unsuitable for agriculture" (with an index of 100)

Table 11.3 Percentage changes in the probability of deforestation per decade relative to base deforestation rate, 1990–2000[a]

Variable	Madagascar	Southwest	Low elevation, East	High elevation	Low elevation, West	Sambirano
Distance to nearest footpath[b]	**−4.6**	−0.8	**−12.6**	**−5.9**	**−6.3**	**8.1**
Distance to nearest road[b]	**−2.7**	**0.8**	**−3.5**	**−3.9**	**−4.7**	**−3.1**
Elevation[c]	**−1.4**	**47.4**	**−2.4**	**−3.6**	**−12.0**	**−5.1**
Inequality[d]	−2.4	**22.4**	−3.7	3.7	**−22.7**	−6.1
Mean monthly per capita household expenditure[e]	**−5.2**	**28.7**	−9.8	**−14.2**	**−9.8**	−3.5
Population density[f]	**1.7**	−5.4	1.0	**3.8**	1.9	5.2
Protected area[g]	**−61.5**	NA[h]	**−59.2**	**−71.8**	**−35.5**	**−67.3**
Slope[i]	**−17.0**	**−78.7**	**−11.1**	**−13.5**	**−15.7**	−1.9
Soil fertility constraints[j]	**6.8**	**13.4**	**9.4**	21.2	−2.4	**23.6**

Boldface indicates a change based on statistical association significant at $P > |z| = 0.01$ (see Tables 11.2 and 11.6)

[a]Base deforestation rates vary for all of Madagascar (8.4%) and for the separate regions of the Southwest (11.9%); Low Elevation, East (9.2%); High Elevation (5.7%), Low Elevation, West (7.4%), and Sambirano (13.2%)

[b]Probability changes based on increase in distance by 1 km

[c]Probability changes based on increase in elevation by 100 m

[d]Probability changes based on increase in Gini coefficient by one standard deviation

[e]Probability changes based on increase in average monthly per capita household expenditure by 100,000 Malagasy francs

[f]Probability changes based on increase in population density by 10 persons per km^2

[g]Probability changes based on placement of protected area where one currently does not exist

[h]Not applicable; insufficient protected areas present in the Southwest region to measure their statistical association with deforestation

[i]Probability changes based on increase in slope by 10°

[j]Probability changes based on increase in soil fertility constraints by 10 – an increase in roughly one constraint category (which ranged from "no constraints" to "unsuitable for agriculture")

increase of 1.7% over the 8.4% base rate (Table 11.3). However, a change in population density of this magnitude would be considerable, given that the mean firaisana-level population density in our sample was about 15 persons per km^2 – meaning that a relatively very large increase in population would yield only a relatively small increase in deforestation.

With regard to income, an increase in mean monthly expenditures of 100,000 Malagasy francs leads to a 0.43% point predicted decrease in the probability of deforestation per decade. Although this represents a 5.2% decline relative to the base rate, it requires a relatively large increase in expenditures to do so – recalling that about three of four rural Malagasy households survive on less than 354,000 Malagasy francs per capita expenditure per month. Reducing income inequality by one standard deviation of the Gini coefficient (see note "e" on Table 11.2 for a definition) increases the likelihood deforestation per decade by 0.2% points, a relative increase of 2.4% above the base deforestation probability. These results on income inequality contrast somewhat with the analysis of income, in that reducing inequality through broad increases in income would yield a slight *increase* in the probability of deforestation.

We found a strong and significant negative association between deforestation and distance from footpaths, distance from roads, and slope – all measures of *accessibility* – suggesting that deforestation is less likely in places more difficult to reach. Note that the spatial resolution of data on transportation access and topography is much finer than for the socioeconomic data examined, providing substantially improved analysis of these variables in the context of deforestation. An increase in distance to footpath or road of 1 km would result in 4.6 and 2.7% reductions in forest loss relative to the base rate. Similarly, a 10° increase in slope results in an absolute decrease of about 1.4% per decade in the probability of deforestation – about 17.0% below the base deforestation rate. In the case of all indicators of access considered, relatively small changes in associated independent variables lead to relatively large predicted changes in the likelihood of deforestation.

In addition to slope, other physical geographic characteristics also showed a statistically significant relationship with deforestation during the 1990s. Elevation has a negative relationship with deforestation. An increase in elevation of 100 m yields a reduction in the absolute probability of deforestation of –0.1% per decade or about 1.4% below the base rate. Higher soil fertility constraints are associated with higher deforestation rates, though the magnitude of the relationship is modest.

One noteworthy finding of our probit analyses, quite reassuring from the conservation standpoint, is that the probability of deforestation is significantly lower inside protected areas compared to areas with similar topography, accessibility, income, and population density but no protection. The multivariate approach taken provides some confidence that the reduced deforestation rate is due to protection itself rather than, for instance, a tendency to place protected areas in inaccessible locations. Other factors held constant, the presence of a protected area will on average decrease the probability of deforestation by 5.2% points per decade. In relative terms, this translates into about 61.5% reduction below the base deforestation rate for Madagascar as a whole.

As noted, the probit model results varied by region. This regional variability reflects differences in physical geography, dominant economic activities, and other characteristics that we used as independent variables in modeling deforestation (Table 11.4). Regional differences in model results also likely are a consequence of highly variable deforestation rates by region during the 1990s (Table 11.5). Statistically, deforestation in the Sambirano and Southwest regions was significantly higher than for the country as a whole during this time period, although because of its small size the impact of the Sambirano on overall deforestation is low.

For most variables, deforestation patterns in the Southwest contrast strongly with patterns found elsewhere in Madagascar (Table 11.6). In this region, higher income and higher income inequality are strongly and significantly (at 0.01, unless

Table 11.4 Mean values for socioeconomic, access, and physical geographic characteristics in forested areas,[a] by geographic region (standard deviations shown in parentheses)

	Southwest	Low elev. East (0–800 m)	Higher elevation (>800 m)	Low elev. West (0–800 m)	Sambirano	All forested areas
Distance to nearest footpath (km)	2.7 (2.2)	3.8 (3.1)	5.1 (3.9)	3.0 (2.4)	5.7 (5.5)	3.7 (3.3)
Distance to nearest road (km)	9.5 (8.1)	15.4 (10.5)	14.8 (9.5)	8.9 (7.3)	11.3 (9.5)	11.9 (9.2)
Elevation (m above sea level)	177 (96)	527 (287)	1,131 (318)	198 (186)	382 (328)	527 (477)
Inequality[b]	0.4 (0.05)	0.04 (0.04)	0.3 (0.06)	0.3 (0.05)	0.3 (0.02)	0.4 (0.05)
Mean monthly per capita household expenditure[c]	236,263 (56,710)	303,483 (72,075)	302,487 (83,060)	336,183 (121,085)	481,057 (206,636)	301,305 (105,116)
Population density (persons/km^2)	15.6 (24.4)	19.8 (18.4)	18.0 (28.8)	7.6 (26.1)	15.2 (23.0)	14.7 (25.9)
Poverty rate (%)[c, d]	82.7 (7.2)	74.7 (10.5)	74.4 (13.6)	67.9 (16.4)	44.4 (10.6)	73.8 (14.7)
Slope (degrees)	1.3 (2.8)	8.0 (9.4)	8.6 (9.4)	2.0 (4.3)	8.7 (9.4)	4.9 (7.8)
Soil fertility constraints	40.2 (16.6)	71.0 (11.4)	71.1 (13.1)	42.2 (17.9)	66.3 (14.2)	55.3 (21.2)

[a]This table only includes forested areas or sample points categorized as forested in 1990. It does not describe the country as a whole

[b]Measured in terms of the Gini coefficient (the ratio between existing income distribution and theoretical income equality across a population). For a perfectly equal distribution, the Gini coefficient equals zero. For complete inequality (where only one person has any income), the Gini coefficient is one

[c]Because of the close association between poverty rate and monthly expenditures, we included only one (the latter) in the statistical analysis of deforestation

[d]Poverty rate is based on a sample headcount and is presented here as percentage of total population in a region or for all forested areas. The poverty line is 354,000 Malagasy francs per capita household expenditures per month – at the 1993 exchange rate with US dollars, about $60

Table 11.5 Forest cover loss (1990–2000) and study localities protected (1990) by geographic region

	Southwest	Low elevation east	Higher elevation	Low elevation west	Sambirano	All forested areas
Percentage of sample points deforested in 2000	11.9	9.2	5.7	7.4	13.2	8.4
Percentage of sample points in a protected area	0.4	4.5	10.7	6.2	5.3	5.9

Table 11.6 Statistical results of probit modeling of deforestation for regions of Madagascar by independent variable

Variable	Marginal effect	Standard error	*Z*-value[a]	$P > \lvert z \rvert$[a]
Southwest				
Distance to nearest footpath[b]	−0.00000101	0.000001	−1.1	0.29
Distance to nearest road[b]	0.00000099	0.0000003	3.7	0.00
Elevation[c]	0.0005646	0.00003	22.3	0.00
Inequality[d]	0.532	0.05	10.5	0.00
Mean per capita household expenditure[e]	0.000342	0.00004	7.9	0.00
Population density[f]	−0.000638	0.0003	−2.4	0.02
Protected area[g]	NA	NA	NA	NA
Slope[h]	−0.00937	0.001	−8.4	0.00
Soil fertility constraints[i]	0.0016	0.0001	12.1	0.00
Low elevation, East				
Distance to nearest footpath[b]	−0.0000116	0.0000009	−12.2	0.00
Distance to nearest road[b]	−0.00000322	0.0000003	−11.9	0.00
Elevation[c]	−0.0000221	0.000008	−2.7	0.01
Inequality[d]	−0.0853	0.07	−1.3	0.21
Mean per capita household expenditure[e]	−0.0000904	0.00004	−2.2	0.03
Population density[f]	0.0000909	0.0001	0.9	0.40
Protected area[g]	−0.0545	0.006	−5.6	0.00
Slope[h]	−0.00102	0.0002	−4.2	0.00
Soil fertility constraints[i]	0.000863	0.0002	4.4	0.00
High elevation				
Distance to nearest footpath[b]	−0.00000337	0.0000004	−7.8	0.00
Distance to nearest road[b]	−0.00000222	0.0000002	−13.0	0.00
Elevation[c]	−0.0000206	0.000005	−4.5	0.00
Inequality[d]	0.035	0.02	1.5	0.14
Mean per capita household expenditure[e]	−0.0000808	0.00002	−4.4	0.00
Population density[f]	0.000214	0.00006	3.6	0.00
Protected area[g]	−0.0409	0.002	−10.7	0.00
Slope[h]	−0.000772	0.0001	−5.9	0.00
Soil fertility constraints[i]	0.00121	0.00009	1.4	0.16
Low elevation, West				
Distance to nearest footpath[b]	−0.00000467	0.0000007	−6.4	0.00

(*continued*)

Table 11.6 (continued)

Variable	Marginal effect	Standard error	*Z*-value[a]	*P* > |z|[a]
Distance to nearest road[b]	−0.00000348	0.0000003	−12.9	0.00
Elevation[c]	−0.0000886	0.00001	−8.1	0.00
Inequality[d]	−0.336	0.03	−9.7	0.00
Mean per capita household expenditure[e]	−0.0000726	0.00001	−5.0	0.00
Population density[f]	0.000138	0.00006	2.4	0.02
Protected area[g]	−0.0263	0.005	−4.1	0.00
Slope[h]	−0.00116	0.0004	−2.6	0.01
Soil fertility constraints[i]	−0.00018	0.00008	−2.2	0.03
Sambirano				
Distance to nearest footpath[b]	0.0000107	0.000001	8.0	0.00
Distance to nearest road[b]	−0.00000411	0.000001	−4.2	0.00
Elevation[c]	−0.0000667	0.00003	−2.6	0.01
Inequality[d]	−0.403	0.4	−1.1	0.29
Mean per capita household expenditure[e]	−0.0000467	0.00003	−1.7	0.09
Population density[f]	0.000686	0.0004	1.9	0.05
Protected area[g]	−0.0888	0.02	−3.3	0.00
Slope[h]	−0.000246	0.0008	−0.3	0.75
Soil fertility constraints[i]	0.00312	0.005	6.2	0.00

[a]*z* and *P* > |z| are the test of the underlying coefficient being 0
[b]Meters
[c]Meters above sea level
[d]Gini coefficient
[e]000s of Malagasy francs
[f]Persons per km^2
[g]Presence/absence; "NA" means "not applicable," in this case because insufficient protected areas existed to enable a statistical analysis
[h]Degrees
[i]Organized as seven constraint categories, ranging from "no constraints" (with an index of 0) to "unsuitable for agriculture" (with an index of 100)

otherwise noted) associated with *higher* rates of deforestation. A 100,000 Malagasy franc increase in income is associated with a 3.4% increase in the decadal deforestation rate – 28.7% higher than the base deforestation rate of 11.9% for this region. A one-standard deviation increase in inequality (see Table 11.4) yields roughly a 2.7% increase in the decadal deforestation rate, about 22.4% above the base rate for 1990–2000. These relationships may reflect the importance of export-oriented maize agriculture in the region's deforestation (Casse et al. 2004). The association may possibly indicate reverse causality, with the higher incomes stemming from profitable deforestation. The likelihood of deforestation increases slightly with remoteness from roads and decreases slightly with population density (significant at 0.02), both unexpected findings. Slope shows a substantial and expected effect – a 10° increase is associated with a 9.4% point decrease in the decadal deforestation probability, about 78.7% below the base rate for the region.

For the remaining four regions, many of the independent variables considered share similar statistical associations with deforestation. The presence of a protected area has a consistently strong negative relationship with forest loss – a likely deterrent whose relative impact ranged from 35.5 to 71.8% compared to the base deforestation rates for the four regions. Slope also emerged as an important variable, associated with lower deforestation rates in four of the five regions examined and likely reflecting accessibility. Proximity to the nearest footpath and to the nearest road, also measuring accessibility, in most cases is strongly and significantly associated with higher deforestation rates.

In contrast, population density had effects of generally modest statistical significance and low magnitude in the remaining four regions. In the Low Elevation West and High Elevation regions, increases in population density by 10 persons per km^2 (which would be substantial in each region) are associated with relatively small increases in deforestation – between 1.9 (significant at 0.02) and 3.5% above regional base rates. The relationship between population density and deforestation is not statistically significant in the Sambirano or in the Low Elevation East.

The association of income with deforestation also is modest. An increase in mean monthly expenditures of 100,000 Malagasy francs in the Low Elevation West and High Elevation regions is associated with a reduction in the decadal deforestation rate of between 9.8 and 14.2%, which themselves are strong marginal effects. However, 100,000 francs represent between 29.8 and 33.1%, respectively, of monthly expenditures in the regions where expenditures are significantly related to deforestation – indicating that although an increase in income can reduce deforestation, that increase must be considerable, at least in relative terms.

11.5 Discussion and Conclusions

In a setting such as Madagascar, where the conservation of irreplaceable biodiversity relies on natural forest that is rapidly disappearing, understanding the variables associated with deforestation is central to developing successful conservation strategies. A rapidly growing rural population mired in poverty whose survival often involves cutting trees for fuel, for timber, or to enable crop production, frequently is proposed in conventional explanations as driving forest loss. But the statistical analyses discussed in this paper do not indicate direct roles for population or poverty in deforestation during the 1990s. Instead, our study reveals a complex relationship between characteristics of the human and physical geography of Madagascar and the removal of forest. Some of these characteristics are sensitive to government policies and development strategies, suggesting possible actions that decision-makers can take to reduce forest loss.

One of the most consistent results of our study is that protected areas appear to reduce the rate of deforestation substantially. Because protected areas may be deliberately sited in locations that are unattractive to agriculture (Cropper et al. 2001), our analysis controlled for observable deforestation risk factors such as remoteness and slope (although not for all factors that might discourage agriculture). Even after these adjustments, the deforestation rate in protected areas generally is about one-fourth to one-half that of nonprotected areas. This is a noteworthy finding, given that the effectiveness of protected areas can vary considerably, especially in low-income countries such as Madagascar (Kistler and Spack 2003; Smith 1997; see Bruner et al. 2001). Notice that we did not explore the differing effect of various types of protected areas – for instance, national parks possibly having a dissimilar impact on forest loss than special reserves. If one type of protected area provides greater protection against deforestation, creating additional protected areas of that type or converting other forms of existing protected areas may improve forest conservation. The apparent success of protected areas in stemming deforestation highlights the importance of former Madagascar president Marc Ravalomanana's 2003 announcement to triple the amount of land in his nation's protected areas by 2012.

Another consistent finding of our study is that increased accessibility often is associated with higher deforestation rates. In this study, we measured accessibility in terms of three variables: proximity to roads, proximity to footpaths, and slope. Although slope is a fixed physical condition, roads and footpaths are not, and their introduction or modification can change access to different areas dramatically. The degree to which roads and footpaths *cause* deforestation remains uncertain, as the placement of both may be due to other reasons (e.g., to provide access to good agricultural land or mining operations) that themselves lead to forest clearing. That stated, because road access often is linked to higher prices of agricultural output and more intensive land use – greater incentives for deforestation ultimately due to greater profitability of commercial farming (see Kistler and Spack 2003) – a causal role may at least in part be the case for roads. Road construction and improvement present major decisions in government policy and development investments, providing a means of controlling access to certain areas to help reduce the rate of forest loss. In contrast to roads, the placement of footpaths generally does not result from policy or development decisions, but rather from decisions of local residents.

Our analyses found weaker statistical association between either population or poverty and deforestation, offering little clear policy guidance. Results indicate that efforts to reduce population density in environmentally sensitive areas would at best very slightly decrease deforestation in those areas. Similarly, the relation between poverty and deforestation is sufficiently weak such that even large increases in income would yield only modest reductions in deforestation. To some extent, both the population and poverty analysis results may be affected by socioeconomic data inadequacies. Recall that both population density and poverty data date to 1993 and are aggregated to the firaisana level. Using information from a single year to model deforestation over a decade, and information averaged across a large geographic unit to model deforestation in a much smaller local unit, undoubtedly affects the measurement of statistical association.

As our analysis shows, deforestation dynamics tend to vary regionally in Madagascar, with those in the Southwest differing markedly from those found elsewhere in the country. Unlike the rest of the island, deforestation in the Southwest is positively associated with higher incomes and levels of inequality and negatively associated with population density. This is consistent with observations that recent deforestation in this region is related in part to export-oriented maize production (Casse et al. 2004). In short, people in the Southwest may have strong financial incentives for deforestation. As a result, the tradeoffs between poverty alleviation and biodiversity conservation likely are steepest in this part of the country. In general, the outcomes of our analyses for the remaining four regions examined are similar to those for all of Madagascar – namely that protected areas tend to reduce deforestation and access via roads and footpaths tend to increase it. However, regional differences in the variables associated with deforestation indicate that reducing deforestation will require strategies tailored for specific regions, if not for particular subregional geographic settings.

The results of this study indicate that addressing what have long been considered the root socioeconomic causes of deforestation in Madagascar, population and poverty, will have limited impacts in reducing the removal of forests. However, there is no reason to expect simple relationships between these variables and forest loss. Population density and income are *joint outcomes* of decisions about where to live, what livelihoods to pursue, and how to pursue them. These decisions are shaped by certain fixed conditions, such as geography and history. They also are shaped by constraints and opportunities that are subject to policy influence – including the location and condition of roads and the placement of protected areas, as well as available agricultural technologies, opportunities for off-farm employment, and so on. Actions along any of these dimensions will have some effect on where people live and their level of income, as well as on patterns of deforestation. In this chapter, we have shown that protected areas and road access, both subject to government policies and development strategies, have strong associations with deforestation. The strategic establishment of protected areas in localities of high conservation priority and limiting the introduction or improvement of roads in environmentally sensitive areas may help reduce local deforestation. A better understanding of how such actions affect other development concerns could promote integrated development strategies that involve sustainable management of their natural resources to conserve biodiversity that also improve the human condition.[2,3]

[2]Joanna Durbin and Bart Minten kindly commented on an earlier version of this paper, helping to improve both clarity and accuracy. Mark Denil recommended key improvements in the design of Figs. 11.1, 11.2, and 11.3.

[3]The findings, interpretations, and conclusions expressed herein are those of the authors and do not necessarily reflect the views of the International Bank for Reconstruction and Development/The World Bank and its affiliated organizations, or those of the Executive Directors of The World Bank or the governments they represent.

References

Abel-Ratovo H, Andrianarison F, Rambeloma T, Razafindraibe R (2000) Analyses des causes raciness socio-économiques de la perte de la biodiversité dans l'ecoregion de forêt tropicale épineuse de Madagascar. World Wide Fund for Nature, Antananarivo

Bertrand A, Sourdat M (1998) Feux et deforestation à Madagascar: Revues bibliographiques. CIRAD, Antananarivo

Brand J, Zurbuchen J (1997) La déforestation et le changement du couvert végétal. In: Un système agro-écologique dominé par le tavy: La région de Beforona, falaise est de Madagascar. Projet Terre-Tany/BEMA, Antananarivo, pp 59–67

Bruner AG, Gullison RE, Rice RE, da Fonseca GAB (2001) Effectiveness of parks in protecting tropical biodiversity. Science 291:125–128

Bureau Central du Recensement (1993) Recensement général de la population et de l'habitat. Direction de la démographie et des statistiques socials, Antananarivo

Cabalzar GP (1990) Opération sauvegarde et aménangment des forêts – côte ouest (SAF-CO Morondava). 1ère phase: Connaisance du milieu. Akon'ny Ala 5:14–21

Cardiff S, Andriamanalina A (2007) Contested spatial coincidence of conservation and mining efforts in Madagascar. Madagascar Conserv Dev 2:28–34

Cardiff S, Befourouack J (2003) The Reserve Speciale d'Ankarana. In: Goodman SM, Benstead JP (eds) The natural history of Madagascar. University of Chicago Press, Chicago, pp 1501–1507

Casse T, Milhøj A, Ranaivoson S, Randriamanarivo JR (2004) Causes of deforestation in southwestern Madagascar: what do we know? For Pol Econ 6:33–48

Chomitz KM, Gray DA (1996) Roads, Land use, and deforestation: a spatial model applied to Belize. World Bank Econ Rev 10(3):487–512

Cropper M, Puri J, Griffiths C (2001) Predicting the location of deforestation: the role of roads and protected areas in north Thailand. Land Econ 77(2):172–186

Dewar RE (1984) Extinctions in Madagascar: the loss of the subfossil fauna. In: Martin PS, Klein RG (eds) Quaternary extinctions. University of Arizona Press, Tucson, AZ, pp 574–593

Dewar RE (1997) Were people responsible for the extinction of Madagascar's subfossils, and how will we ever know? In: Goodman SM, Patterson BD (eds) Natural change and human impact in Madagascar. Smithsonian Institution Press, Washington, DC, pp 364–377

Dewar RE (2003) Relationship between human ecological pressure and the vertebrate extinction. In: Goodman SM, Benstead JP (eds) The natural history of Madagascar. University of Chicago Press, Chicago, pp 119–122

Donque G (1972) The climatology of Madagascar. In: Battistini R, Richard-Vindard G (eds) Biogeography and ecology of Madagascar. W. Junk, The Hague, pp 87–144

Duffy R (2005) Global environmental governance and the challenge of shadow states: the impact of illicit sapphire mining in Madagascar. Dev Change 36:825–843

Dufils JM (2003) Remaining forest cover. In: Goodman SM, Benstead JP (eds) The natural history of Madagascar. University of Chicago Press, Chicago, pp 88–96

Durbin J, Bernard K, Fenn M (2003) The role of socioeconomic factors in loss of Malagasy biodiversity. In: Goodman SM, Benstead JP (eds) The natural history of Madagascar. University of Chicago Press, Chicago, pp 142–146

Economist Intelligence Unit (1998) Madagascar country profile, 1998–99. Economist Intelligence Unit, London

Erdmann TK (2003) The dilemma of reducing shifting cultivation. In: Goodman SM, Benstead JP (eds) The natural history of Madagascar. University of Chicago Press, Chicago, pp 134–139

Fischer G, van Velthuizen H, Shah M, Nachtergaele FO (2002) Global agro-ecological assessment for agriculture in the 21st century: methodology and results. International Institute of Applied Systems Analysis, Laxenburg, Austria

Freudenberger KS (2003) The Fianarantsoa-East Coast Railroad and its role in eastern forest conservation. In: Goodman SM, Benstead JP (eds) The natural history of Madagascar. University of Chicago Press, Chicago, pp 139–142
Gade D (1996) Deforestation and its effects in highland Madagascar. Mt Res Dev 16:101–116
Gautier L, Goodman SM (2003) Introduction to the flora of Madagascar. In: Goodman SM, Benstead JP (eds) The natural history of Madagascar. University of Chicago Press, Chicago, pp 229–250
Goodman SM, Benstead JP (eds) (2003) The natural history of Madagascar. University of Chicago Press, Chicago
Goodman SM, Rakotondravony D (2000) The effects of forest fragmentation and isolation on insectivorous small animals (Lipotyphla) on the central high plateau of Madagascar. J Zool 250:193–200
Goodman SM, Ganzhorn JU, Rakotondravony D (2003) Introduction to the mammals. In: Goodman SM, Benstead JP (eds) The natural history of Madagascar. University of Chicago Press, Chicago, pp 1159–1186
Green GM, Sussman RW (1990) Deforestation history of the eastern rain forests of Madagascar from satellite images. Science 248:212–215
Hance WA (1975) The geography of modern Africa, 2nd edn. Columbia University Press, New York
Harper G (n.d.) Madagascar: forest cover and deforestation, 1990–2000. Science and Knowledge Division, Conservation International, Arlington, VA (unpublished manuscript on file)
Harper G, Steininger M, Tucker C, Juhn D, Hawkins F (2007) Fifty years of deforestation and forest fragmentation in Madagascar. Environ Conserv 34:325–333
Helfert MR, Wood CA (1986) Shuttle photos show Madagascar erosion. Geotimes 31:4–5
Horning NR (2003) How rules affect conservation outcomes. In: Goodman SM, Benstead JP (eds) The natural history of Madagascar. University of Chicago Press, Chicago, pp 146–153
Jarosz L (1993) Defining and explaining tropical deforestation: shifting cultivation and population growth in colonial Madagascar (1896–1940). Econ Geogr 64:366–379
Jolly A (1989) The Madagascar challenge: human needs and fragile ecosystems. In: Leonard HJ (ed) Environment and the poor: development strategies for a common agenda. Transaction Books, New Brunswick, NJ, pp 189–211
Jolly A, Oberlé P, Albignac R (1984) Key environments: Madagascar. Pergamon, Oxford, UK
Kistler P, Spack S (2003) Comparing agricultural systems in two areas of Madagascar. In: Goodman SM, Benstead JP (eds) The natural history of Madagascar. University of Chicago Press, Chicago, pp 123–134
Klein J (2002) Deforestation in the Madagascar highlands – Established "truth" and scientific uncertainty. GeoJournal 56:191–199
Kull CA (2003) Fire and the management of highland vegetation. In: Goodman SM, Benstead JP (eds) The natural history of Madagascar. University of Chicago Press, Chicago, pp 153–157
Langrand O, Wilmé L (1997) Effects of forest fragmentation on extinction patterns of the endemic avifauna on the central high plateau of Madagascar. In: Goodman SM, Patterson BD (eds) Natural change and human impact in Madagascar. Smithsonian Institution Press, Washington, DC, pp 280–305
Laub-Fischer R, Wehr R, Roge A (1997) L'approche planification de la gestion du terroir est-elle un remède efficace dans la gestion durable des resources naturelles? Akon'ny Ala 22/23:34–49
McConnell W (2002) Madagascar: Emerald isle or paradise lost? Environment 44:10–22
Messerli P (2002) Le dilemme entre la culture sur brûlis et la conservation des resources naturelles sur la Falaise Est de Madagascar. Alternatives et améliorations du système agro écologique en vue d'un dévelopment plus durable. Thése de doctorat, Centre pour le Développmement et l'Environnement, Institut de Géographie, Universite de Berne, Bern
Ministère du Développement Rural et de la Réforme Agraire (1995) Le système national de vulgarization agricole. Rapport juin 1995. Ministère du Développement Rural et de la Réforme Agraire, Antananarivo

Mistiaen J, Özler B, Razafimanantena T, Razafindravonona J (2002) Putting welfare on the map in Madagascar. Africa Region Working Paper Series No. 34. World Bank, Washington, DC
Mittermeier RA, Tattersall I, Konstant WR, Meyers DM, Mast RB (1994) Lemurs of Madagascar. Conservation International, Washington, DC
Mittermeier RA, Konstant WR, Mittermeier CG, Mast RB, Murdoch JD (1999) Madagascar. In: Mittermeier RA, Myers N, Robles Gil P, Mitermeier CG (comilers) Hotspots. CEMEX, Mexico City, pp 188–200
Mittermeier RA, Langrand O, Lowry PP, Schatz G, Gerlach J, Goodman S, Steininger M, Hawkins F, Raminosoa N, Ramilijaona O, Andriamaro L, Randrianasolo H, Rabarison HL, Rakotobe ZL (2004) Madagascar and the Indian Ocean Islands. In: Mittermeier RA, Robles Gil P, Hoffman M, Pilgrim J, Mittermeier CG, Lamoreux J, Da Fonseca GAB (compilers) Hotspots revisited. CEMEX, Mexico City, pp 138–144
Morris P, Hawkins F (1998) Birds of Madagascar, a photographic guide. Yale University Press, New Haven, CT
Myers N (1986) Tropical deforestation and a mega-extinction spasm. In: Soulé M (ed) Conservation biology: the science of scarcity and diversity. Sinauer Associates, Sunderland, MA, pp 394–409
Myers N, Mittermeier RA, Mittermeier CG, da Fonseca GAB, Kent J (2000) Biodiversity hotspots for conservation priorities. Nature 403:853–858
Population Reference Bureau (2004) 2004 world population data sheet of the Population Reference Bureau. Population Reference Bureau, Washington, DC
Preston-Mafham K (1991) Madagascar. A natural history. Facts on File, New York
Rakotomamonjy J (2000) Utilisation de deux images Landsat et du SIG pour la cartographie de la dégradation forestière en fonction de la distance de la route nationale 2 et du chemin de fer Tananarive Côte Est. Diplôme d'Etudes Supérieures Specialisées (réalisée au CFSIGE), Ecole Superieure Polytecnique, Université d'Antananarivo, Antananarivo
Rakotomanana JL (1989) La conservation des sols – côté paysans. Akon'ny Ala 3:15–19
Rakotonindrina R (1988) Développment intégré en zones forestiers. In: Maldague M, Matuka K, Albignac R (1988) Actes du séminaire international dur la gestion de l'environment, zone africaie de l'Océan Indien. UNESCO, Paris, pp 245–252
Ratsirarson J (2003) Réserve Spéciale de Beza Mahafaly. In: Goodman SM, Benstead JP (eds) The natural history of Madagascar. University of Chicago Press, Chicago, pp 1520–1525
Sarrasin B (2006) The mining industry and the regulatory framework in Madagascar: some developmental and environmental issues. J Clean Prod 14:388–396
Schatz GE (2000) Endemism in the Malagasy tree flora. In: Lourence WR, Goodman SM (eds) Biogeography of Madagascar. Memoires de la Societe de Biogeographie, Paris, pp 1–9
Smith AP (1997) Deforestation, fragmentation, and reserve design in western Madagascar. In: Laurence W Jr (ed) Tropical forest remnants: ecology, management, and conservation of fragmented communities. University of Chicago Press, Chicago, pp 415–441
Smith AG, Smith DG, Funnell BM (1994) Atlas of mesozoic and cenozoic coastlines. Cambridge University Press, Cambridge, UK
Smith AP, Horning N, Moore D (1997) Regional biodiversity planning and lemur conservation with GIS in western Madagascar. Conserv Biol 11(2):498–512
Sussman RW, Green GM, Sussman L (1994) Satellite imagery, human ecology, anthropology, and deforestation in Madagascar. Hum Ecol 22(3):333–354
United Nations Population Division (2005) World population prospects: The 2004 revision population database; 2005. http://esa.un.org/unpp. Accessed 02 June 2005
Vallan D (2000) Influence of forest fragmentation on amphibian diversity in the nature reserve of Ambohitantely, highland Madagascar. Biol Conserv 96(1):31–43
Vincelette M, Randrihasipara L, Ramanamanjato J-B, Lowry PP II, Ganzhorn JU (2003) Mining and environmental conservation: the case of QIT Madagascar Minerals in the southeast. In: Goodman SM, Benstead JP (eds) The natural history of Madagascar. University of Chicago Press, Chicago, pp 1535–1537

Walsh A (2004) In the wake of things: speculating in and about sapphires in northern Madagascar. Am Anthropol 106:225–234

World Bank (2003) Madagascar. Rural and economic sector review, vol 1: Main report. World Bank, Washington, DC

World Bank (2005) Little Data Book 2005. World Bank, Washington DC

Wright PC (1997) The future of biodiversity in Madagascar. A view from Ranomafana National Park. In: Goodman SM, Patterson BD (eds) Natural change and human impact in Madagascar. Smithsonian Institution Press, Washington, DC, pp 381–405

Zeiler M (1999) Modeling our world: the ESRI® guide to geodatabase design. ESRI, Redlands

Part III
Perspectives on Human Interactions with Biological Diversity

Chapter 12
A Coupled Natural and Human Systems Approach Toward Biodiversity: Reflections from Social Scientists

Thomas W. Crawford and Deirdre Mageean

12.1 Introduction

The topic of this book – human population's impacts on biodiversity – presents significant challenges to the research community due to its inherent integrative nature (Covich 2000). Integration typifies the recent currency of coupled natural and human systems as a research frontier across the natural and social sciences (Pickett et al. 2005; Liu et al. 2007). In this essay, we discuss and critique the key themes and findings presented in the preceding chapters written by a diverse array of social and natural scientists. We do so from vantage points as social scientists, trained specifically as human geographers and affiliated with a branch of geography that focuses on the study of patterns and processes shaping human interactions with the environment. Our geographic tradition encompasses human, political, cultural, social, and economic aspects concerning the causes and consequences of the spatial distribution of human activities. We also share a common research interest within the more specialized fields of demography and population geography. Importantly with respect to themes in this volume, our research has significant connections with geographic information science – specifically, spatially explicit approaches and analyses involving linkages of human and environmental data. While cognizant of and even engaged with natural science literatures and collaborators, we do not personally engage in field-based biological or ecological research concerning biodiversity.

Having laid out our personal contexts, it is important to convey what we understand biodiversity to mean. At its most simple, biodiversity is the variation of life at all levels of biological organization. Biodiversity can be further defined in terms of genetic diversity, species diversity, and ecosystem diversity (Table 12.1). We note that the contributions in this volume focus either on species or ecosystem diversity. Several metrics are employed by authors to quantify aspects of biodiversity. Some of the chapters in this volume discuss and review literatures that investigate these differing definitions using the varied empirical metrics. An alternative is the use of the number of threatened species which can be viewed as an indirect measure. Other chapters do not engage directly with actual biodiversity measures and instead focus on describing and measuring spatial distributions of

R.P. Cincotta and L.J. Gorenflo (eds.), *Human Population: Its Influences on Biological Diversity*, Ecological Studies 214,
DOI 10.1007/978-3-642-16707-2_12,

Table 12.1 Biodiversity definitions and metrics

Definitions	
Genetic diversity	Diversity of genes within a species
Species diversity	Diversity of species in an ecosystem
Ecosystem diversity	Diversity at a higher level of organization – the ecosystem. Biological "hotspots" are prime examples of high ecosystem diversity
Metrics	
Species richness	The number of species in a given area
Species abundance	The evenness of distribution of individuals among species in a community
Simpson index	The probability that two randomly selected individuals in a system belong to the same species
Shannon index	The information entropy of the species distribution. It accounts for both species richness and evenness
Alpha diversity	Identical to species richness
Beta diversity	Species diversity between ecosystems. It compares the number of species unique to each ecosystem
Gamma diversity	A measure of the overall diversity for the different ecosystems with a region

human population and/or land use and land cover change. Both of these foci also can be considered as indirect indicators of human patterns and processes that have implications for biodiversity.

As discussed in Chap. 1, the status of population–environment inquiries is arguably underdeveloped due to the disciplinary boundaries that so often separate the biological and social sciences. The chapters in this volume represent a welcome addition and enhancement to the field due to their collective attempts to soften these boundaries and promote a bona fide coupled natural–human systems science approach toward biodiversity. However, at the most basic level, the outcome of interest is biodiversity – or biodiversity loss. Adopting a regression-based formulation for the sake of discussion, biodiversity is clearly positioned as a dependent variable to be explained or predicted by a theoretically informed and comprehensive set of independent variables. Notwithstanding measurement difficulties, particularly at varying time and space scales, it is clearly the role of biological collaborators to develop appropriate biodiversity datasets and measures. Looking to the other side of the equation, Edward Wilson suggests a set of factors denoted by the acronym HIPPO that threaten biodiversity: habitat destruction, invasive species, pollution, population, and overharvesting (Wilson 2002, 2005). HIPPO provides a short-hand indication of where, for selected factors, social scientists are well positioned to contribute on interdisciplinary research teams. It is notable that in many of this volume's chapters, explicit biodiversity measures are largely absent. Other chapters focus more explicitly on selected HIPPO factors, particularly population and habitat (e.g., land use and cover), in a stand-alone fashion. Stand-alone research of both biodiversity measures and HIPPO is indeed warranted. However, significant challenges remain in order to investigate "both sides of the equation" in a more integrative fashion.

12.2 Human Dimensions of Biodiversity as a Grand Challenge

In 2001, the US National Research Council (NRC) published the report Grand *Challenges in Environmental Sciences* (NRC 2001). Contributors were charged with the task of defining the most important challenges that multidisciplinary research is best positioned to tackle during the twenty-first century. Of the eight challenges (Table 12.2), it is noteworthy that both biological diversity and land use dynamics were identified, and that both were among the four challenges recommended for immediate research investments. While our focus is on biological diversity, we find the inclusion of land use dynamics to be particularly salient because we perceive human drivers of land use dynamics also to be hypothesized drivers of biodiversity change. Additionally, multidisciplinary research on land use/cover change has progressed substantially since the 1990s and may offer at least a partial template that can help guide biodiversity research.

While land use is one of the more prominent themes relevant to biodiversity loss, the NRC's biodiversity grand challenge intended to incorporate a broad array of human dimensions more inclusively to help guide policy solutions. Gorenflo and Brandon (2006) identify and analyze a set of human dimensions – (1) human demographics, (2) land use and land cover, and (3) agricultural suitability – to evaluate the feasibility of prioritizing a conservation network of global biodiversity sites. While more detail and nuance would be required for similar analyses at finer geographic scales, Gorenflo and Brandon's method provides an approach to evaluating feasibility on a large scale – a necessity in most biodiversity policy environments, where limited financial resources are available.

For each challenge, the NRC report included a set of *indicators of scientific readiness* and a list of *important areas for research*. Here, we use these items as guideposts to discuss how the chapters in this volume fit within important streams of biodiversity research. Table 12.3 lists the NRC indicators of scientific readiness (NRC 2001), a categorical heading that we slightly modify and call *scientific themes*. We connect to these themes via common numeric labels the NRC-defined *important areas for research*. We use this as a framework to map the volume chapters to one or more of the connected scientific themes and important areas for research, in effect presenting an informal meta-analysis (Table 12.4). Aspects of this conceptual mapping merit clarification. In many cases, the connections that we

Table 12.2 Grand challenges in environmental sciences

1. Biogeochemical cycles
2. Biological diversity and ecosystem functioning[a]
3. Climate variability
4. Hydrologic forecasting[a]
5. Infectious disease and the environment[a]
6. Institutions and resource use
7. Land use dynamics[a]
8. Reinventing the use of materials

Adapted from NRC (2001)
[a]Recommended immediate research investments

Table 12.3 Biodiversity scientific themes and important areas for research

Scientific theme	Important areas for research connected to scientific theme
(1) Advances in understanding biogeography, speciation, and extinction	(1) Produce a quantitative, process-based theory of biological diversity at the largest possible variety of spatial and temporal scales
(2) Progress in understanding the interaction of biodiversity and ecosystem functioning	(2) Elucidate the relationship between diversity and ecosystem functioning
(3) New and improved tools	(3) Improve tools for rapid assessment of diversity at all scales
(4) Progress in conservation science	(4) and (5) Develop and test techniques for modifying, creating, and managing habitats that can sustain biological diversity, as well as people and their activities
(5) Integration of ecology with economics, psychology, and sociology	

Adapted from NRC (2001)

draw are fairly strong. In other cases, connections are weaker. For example, we see many instances of new and improved tools, Theme (3). What constitutes new and improved tools can be debated. Geographic information system (GIS) technology is no longer a new tool; however, innovative GIS applications involving new or existing data are mapped to this theme. Additionally, chapters not dealing directly with conservation design or habitat management but whose findings have implications for these issues are mapped to Theme (4). Lastly, we view the intent of Theme (5) to be to integrate ecology with the social and behavioral sciences or the human sciences in general. While we leave this theme as stated in its original NRC form, we suspect that others would be comfortable including disciplines such as anthropology, geography, and political science.

12.3 Meta-Analysis Results

Results of our meta-analysis show that the chapters span a diverse array of the identified themes and operate across multiple scales (see Table 12.4). Given the volume's focus on human dimensions, one should not expect that every contribution cover every aspect of biodiversity. It is notable, however, that two chapters (Chaps. 5 and 7) engaged to at least some extent with all five themes. Chapter 5 was clearly the most theory-oriented and asked what theories from ecology or the social sciences can be used or modified to examine human impacts on biodiversity. Drawing from biological theories of allometric growth and social science theories of Malthus and Boserup, it yielded quantitative theoretical models and four testable hypotheses regarding intensification's effect on biodiversity: *nutrient theft*, *successional retreat*, *anthropogenic bias*, and *institutional bias*. Chapter 7 was the sole contribution to investigate aquatic biodiversity explicitly. While less theoretical than

Table 12.4 Meta-analysis summary

Chapter	Title	Themes	Geospatial	Spatial extent (resolution)	Temporal scale	Main contribution
2	Projecting a gridded population of the world using ratio methods of trend extrapolation	3, 5	Yes	Global (5 km)	1990–2025	New method to map projected population distributions
3	Physical environment and the spatial distribution of human population	3, 4, 5	Yes	Global (2.5 km)	2000	Problem of coastal, urban population in the tropics
4	Behavioral mediators of the human population effect on global biodiversity losses	5	No	NA	2 mya – present	Identifies strong positive correlation between population density and threatened species
5	The biological diversity that is humanly possible: three models relevant to human population's relationship with native species	1, 2, 3, 4, 5	No	NA	NA	Theory development and testable hypothesis regarding intensification
6	Biodiversity on the urban landscape	1, 2, 5	No	NA	Late twentieth century	Review of theory and empirical research; role of exotic species
7	Indicators for assessing threats to freshwater biodiversity from humans and human-shaped landscapes	1, 2, 3, 4, 5	Yes	Global, continental, river basin	Late twentieth century	Geospatial methods to identify aquatic threats at multiple scales
8	Cross-cultural analysis of human impacts on the rainforest environment in Ecuador	3, 4, 5	Yes	Subregion (village, household)	Late twentieth century	Human and cultural ecology approach at community and household scales
9	Human demography and conservation in the Apache Highlands ecoregion, US–Mexico borderlands	3, 4, 5	Yes	Ecoregion (census blockgroup, AGEB)	1500 BP – present	Detailed account of demographic history and its implications for conservation
10	Long-term ecological effects of demographic and socioeconomic factors in Wolong Nature Reserve (China)	3, 4, 5	Yes	Reserve – 2,000 km^2	1996–2026	Agent-based modeling of demographic effects on reserve habitat
11	Exploring the association between people and deforestation in Madagascar	3, 4, 5	Yes	National (1 km)	1990–2000	Socio-economic, spatial, and environmental drivers of deforestation

Chap. 5, its review demonstrated advances in understanding aquatic biogeography, threats to biodiversity, and ecosystem functioning – particularly linkages between terrestrial and aquatic environments. Its main contribution was to review approaches to assessing threats using integrative geospatial data sets at multiple scales.

In terms of theme frequency, the modal theme was Theme (5): integration of ecology with the human sciences. This theme was present to varying degrees in every contribution. In some cases, this was reflected by the incorporation of basic demographic concepts and measures such as population size, growth, and density. Stronger cases engaged more deeply with demographic issues such as fertility, mortality, natural increase, migration, marital status, age structure, the household life cycle, education, and labor pools (see Chaps. 8–10). To investigate economic behavioral factors, authors integrated measures such as gross national product, household expenditures, income inequality, fuelwood demand, and energy prices (see Chaps. 4, 8–11) – albeit at varying units of aggregation. Theme (3), new and improved tools, had the second highest frequency. As a caveat, our interpretation is almost totally based on the use of geospatial data and technologies such as GIS, remote sensing, and GPS. While not new in a strict sense, the use of geospatial approaches represents the currency evident in much global change research regarding the linkage of "pixels to people" (Liverman et al. 1998). Linkages evident in this volume range in scale from abstract gridded spaces at the global scale (Chaps. 2 and 4) to individual household plots at the local scale (Chap. 8). A geospatial paradigm is clearly evident in biodiversity research. Theme (4), progress in conservation science, had the third highest frequency, although the theme was largely couched in terms of the implications for conservation rather than direct testing of the impacts of conservation management techniques on biodiversity. Themes (1) and (2) were the least evident. These themes relate most strongly to basic and applied ecological science. Their low frequencies are most likely due to what we perceive to be a heavier concentration of social scientists among the authors, a fact that might be expected given the overarching theme of human dimensions. Despite the lower quantity, these chapters (Chaps. 5–7) present high quality discussions of theoretical and empirical research regarding threats to biodiversity and how they relate to ecosystem functions and potential human drivers.

Looking at how themes tended to cluster, it is noteworthy that Themes (3), (4), and (5) were copresent in over half of the chapters. This may suggest an emerging integrative paradigm among social and ecological scientists that investigates conservation science using geospatial and other techniques. If this is the case, then, such a program would likely benefit by an even stronger coupling with ecological scientists than is reflected in this volume.

Central to any geographic perspective on issues of land use change, or population impacts on biodiversity, is the issue of the appropriate scale of analysis and, consequently, appropriate methods. This has been a concern from some of the earliest work on the human impact on the environment, through the "socializing of the pixel," to the authors of the chapters in this book and one which confronts attempts to model the links between population levels (and rates of change) to biodiversity loss. As noted in the introduction, the breadth of work here emerges as analysis at three geographic scales: the global, the biome, and the local ecosystem. The early chapters (Chaps. 2–5) employ a global scale (see Table 12.4). Others (Chaps. 7 and 8) employ multiple scales, while the remaining chapters focus on a single region.

One of the principal problems bedeviling attempts to integrate ecology with human sciences has been the incompatibility of geo-spatial units used in the respective disciplines. Rarely, if ever, does the extent of a major ecosystem coincide with political boundaries or other geographic units used to record human population and other social science data. A second, major challenge is choosing the appropriate scale to employ in analysis.

At issue here is the fact that results of statistical analyses almost always depend on the spatial scale of the analysis, and a variable which is identified as being significant at one scale may be insignificant at another (Plane and Rogerson 1994). For instance, a straightforward link between population levels (or rates of change) and deforestation, demonstrable at a large scale, might disappear in place-specific analyses. Thus, what appear to be significant correlations may be spurious. This, as Geoghegan points out, is "one of the major scaling issues that confront modeling efforts" (Geoghehan 1998). Such issues particularly beset those studies of population growth and/or urbanization at a global scale. As Harte (2007) has detailed recently, such large-scale studies also have to deal with the fact that the environmental consequences of increasing population size tend to be dynamic and non-linear. Thus, the assumption that environmental degradation grows in proportion to population size or increases given fixed consumption and production is misplaced. Studies of urbanization at the international or global level have to grapple with the crude scale at which population densities and urbanization are calculated. Measures, such as population density, are highly sensitive to geographic scale and can, therefore, be quite crude or misleading (Rain et al. 2007). As those authors point out, the use of distributions of local population densities (now increasingly possible because of global remote sensing and GIS) offers greater potential for understanding the ecological effects of urbanization.

A number of the chapter authors, such as Abell et al. (Chap. 7), recognize that population density is a coarse proxy for specific indicators of ecological disturbance, while Chaps. 4, 8, and 11, in particular, point to the complex ways in which other behavioral, socio-economic and infrastructural variables interact with population variables to impact biodiversity. The results obtained by the authors point to the need to employ and integrate analyses at different scales.

12.4 Connections with Land Change Science

The problem of land use dynamics is another grand challenge identified by the NRC. Interestingly, substantial portions of this volume examine land use/cover change. Recalling Wilson's HIPPO acronym, this should not be surprising given the role that land change plays in altering habitats (the "H") that can directly impact biodiversity. Interactions of human population with land change are also important since terrestrial (and aquatic) change accompanying humans located in geographical space is arguably the proximate driver of biodiversity loss. It is, therefore, worthwhile to consider the trajectory of land change science, similarities this challenge has with the biodiversity challenge, and insights that it may provide for biodiversity research.

During the 1990s, concern regarding environmental problems associated with human-driven land use/cover change led the international research community to establish the Land Use and Cover Change (LUCC) program – a joint program of the International Human Dimensions Program on Global Environmental Change (IHDP) and the International Geosphere–Biosphere Program (IGBP). Much of the impetus behind LUCC came from prior and emerging work using remote sensing analyses that inventoried, mapped, and documented land change from local to global scales. Hotspots of deforestation along with agricultural intensification and urbanization were major themes in much of this research. Three interlocking strategies summarized the LUCC science implementation plan (Table 12.5) (Lambin et al. 1999). Case studies resulting from these strategies frequently employ geospatial strategies involving remote sensing and GIS, the spatial linkage of human populations to impacted landscapes, and theory development regarding human drivers of land use/cover change (for summaries and representative examples, see Liverman et al. 1998; Walsh and Crews-Meyer 2002; Entwisle and Stern 2005).

A generalized theoretical framework undergirding much of land change science posits the influence of proximate and underlying causes of change as mediated by exogenous factors (Fig. 12.1). Descriptive and analytical approaches involving remotely sensed change detection, regression techniques, and more recently, dynamic spatial simulation (e.g., cellular automata and agent-based models) characterize methodological strategies. Intensive fieldwork involving household surveys and ethnographic methods to quantify and interpret human actors responsible for landscape change also figure prominently.

Selected problematic issues confronting the development of land change science (Rindfuss et al. 2004) are germane to integrative biodiversity research. For example, *data aggregation* can lead to risk of committing the ecological fallacy – an

Table 12.5 Land use and cover change science implementation plan

Strategies	Contributing fields
(1) Development of case studies to analyze and model the processes of land use change and land management in a range of generalized global situations	Human ecology, land economics, demography, history
(2) Development of empirical, diagnostic models of land cover change through direct observations and measurements of explanatory factors	Remote sensing, GIS, spatial modeling
(3) Utilization of analysis from (1) and (2) for the development of integrated prognostic regional and global models	Economic modeling, integrated assessment

Adapted from Lambin et al. (1999)

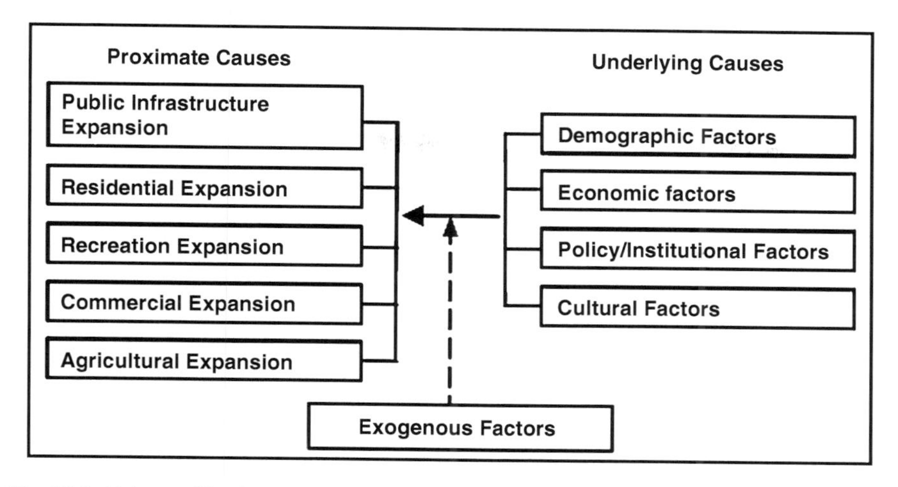

Fig. 12.1 Drivers of land use and land cover change

error in inference demonstrated in classic social demographic research by Robinson (1950). This relates directly to the issue of scale. Chapter 4 demonstrated a strong association between population density and the number of threatened species at the country level. As a first order of approximation, it seems clear that levels of human density do have negative impacts. However, does this relationship hold at differing aggregation levels? More importantly, one can argue that countries do not directly engage in behavior effecting biodiversity, but that agents such as households, firms, agencies, or governing bodies wield the critical behaviors and drivers. Alternatively, there are situations where national policies have implications for biodiversity; for example, foreign policy, military interventions, tourism and natural resource policies, agricultural policies, transportation policies, and alternative energy policies. Researchers should never simply take scale for granted but instead must ask what scale(s) is most appropriate for a particular study? Most likely there is no single answer and research operating at multiple scales can best lead to new understandings and appropriate policy interventions.

Given the importance of geospatial technologies in both land change and biodiversity research, the challenge of *linking people to pixels* is another formidable challenge. Land change researchers have demonstrated that it is possible, with substantial effort, to link individual household parcels and/or agricultural plots to households and villages responsible for their management (Crawford 2002; Evans and Moran 2002; Rindfuss et al. 2002; Fox et al. 2003). As Chap. 6 demonstrates, there are challenges in enlisting property owners to participate in biodiversity studies. It is also possible at broader scales, such as conservation reserves or large publicly owned lands, to delimit areas spatially and enable such linkages. More problematic is the increasing presence of distant and seemingly vague linkages that are connected to globalization. Household ecological footprints in most developed countries range far and wide, drawing on agricultural and industrial lands across the globe each with their respective biodiversity endowments and issues. Tourism also figures in to the issue of far reaching footprints. Future research should strive to make these links more explicit.

Data quality and measurement issues present their own unique challenges. Data quality for population data and other human data vary. Generally speaking, datasets from censuses tend to be highly reliable, while global datasets tend to provide reliable information at large scales and information of variable reliability for small areas. As stated earlier, for biodiversity these topics are clearly within the purview of biological scientists, and here we must admit a certain ignorance of terrain that may be well covered within the biological literature. We are more familiar with these issues in land change science where standard protocols for accuracy assessment exist for data products derived from remotely sensed imagery. Beyond quantitative accuracies of final products, there is the question of classification schemes which influence what one is measuring. For example, a generalized "urban" land cover in fact encompasses many different urban land uses. In short, again, measurement scales matter. Regarding biodiversity, and again speaking as social scientists, we wonder what specific biodiversity measures (e.g., richness, diversity, number threatened) are most important. We suspect that measures will vary in importance depending on the nature of questions asked.

Spatial scientists, including many land change scientists, grapple with the problem of *spatial autocorrelation* when performing inferential analyses. Positive spatial autocorrelation exists where measured attributes of nearby observations tend to be more similar than distant observations, a situation that is common to many datasets. Spatial autocorrelation is also known as spatial dependence and presents a problem for inferential analyses due to violation of the assumption of independent data observations. Potential solutions include sampling or survey designs that are spaced beyond lag distances at which spatial autocorrelation exists. Examination of autocorrelation structure via semivariograms derived from test areas can help determine appropriate lags. A priori knowledge of a region's autocorrelation structure may serve as a substitute to guide strategies. Alternatively, for regression techniques such as those in Chaps. 4, 9, and 11, relatively new methods exist that explicitly account for the presence of autocorrelation. For example, GeoDa is a spatial lag and spatial error regression modeling software available for free web download (Anselin et al. 2006). Geographically weighted regression is another recent method, available in GWR software, that enables spatially varying and mapable slope coefficients (Fotheringham et al. 2002). Methods to handle temporal autocorrelation (i.e., serial correlation) have a much longer history and are well developed. Space–time analysis of regional systems software has been recently developed to examine the effects of both space and time (Rey and Janikas 2006).

Data aggregation, *linking people to pixels*, *data quality and measurement*, and *spatial autocorrelation* form a selection of challenges shared by both the land change and biodiversity science communities. Our insertion of land change science into the discussion is not meant to imply that biodiversity research is somehow deficient in comparison. In spite of significant advances, land change science continues with efforts to overcome these challenges. Importantly though, the land change science community has crystallized to the point where a discourse framed around "best practices" now exists to promote comparable and generalizable research (Rindfuss et al. 2004). Fundamental to both communities is their focus on coupled natural–human systems. We find the similarities in challenges that both communities face to be striking and suggest that, given the clear relevance of land change to biodiversity, these two communities can benefit from a shared knowledge base and future collaborations.

12.5 Neglected Human Dimensions?

The chapters in this volume demonstrate well the multidimensional character of the study of population effects on biodiversity. Joining the approaches of ecologists, demographers, and scientists increases analytical capacity while simultaneously revealing methodological challenges and the need for the consideration and integration of other "human dimensions."

In a paper on neglected dimensions of global land use change, Gerhard Heilig (1994) questioned the conventional approach in studying land use changes "in which agriculture-related alterations are viewed as driven by population growth" and pointed to the need to consider the effects of lifestyles, food preferences, manmade catastrophes, armed conflict, urban infrastructure expansion, industrial production, fossil resource exploration, and transportation. To this we might add political systems as well as public policies (which can prevent, protect, or mitigate) and the socio-economic systems which they reflect. Of important note here is the growing literature on the relationship between population, poverty, and environmental degradation (Blaikie and Brookfield 1987; Duraiappah 1998). Finally, the consideration of sociocultural constructions, different societies' perceptions of risk and vulnerability, and their understanding of place and space are crucial yet not fully integrated into current analyses (Marandola and Hogan 2006).

As a number of chapters in this volume demonstrate, we have advanced in our analyses from a simplistic, causal sequence from population growth leading to increased loss of biodiversity. Evidence from this volume, both empirical studies and the literature reviewed, suggests that human dimensions have not been neglected, and that the research community has indeed responded in positive ways to the NRC's grand challenge regarding biodiversity. However, much remains in bringing together the approaches of biophysical and socio-demographic scientists.

12.6 Conclusion

Biocomplexity (Colwell 1998; Michener et al. 2001; Dybas 2001) has emerged as an important perspective that connects to many of the issues regarding biodiversity presented in this volume. Drawing from research on complex systems (i.e., "complexity theory"), biocomplexity has been defined as "the properties emerging from the interplay of behavioral, biological, chemical, physical, and social interactions that affect, sustain, or are modified by organisms, including humans" (Michener et al. 2001, p. 226). Complex systems, such as ecosystems, landscapes, and biomes, are composed of large numbers of interacting agents (e.g., human population, firms, organisms, species, and energy and nutrient flows) located in space and time that interact across multiple scales in ways that generate a trajectory of system behavior often characterized by feedbacks, nonlinearities, phase transitions, path dependence, and emergence. Such systems are complex in that they typically are not amenable to reductionist scientific strategies and instead often require multidisciplinary teams and frameworks to describe and understand system behavior.

As a hypothetical example, incremental additions of human population via in-migration and their local behaviors and interactions within a regional system may reach a critical threshold whereby receiving regions undergo a phase transition leading to the emergence of new modes and patterns of regional social interactions (integration vs. conflict), economic vitalities (traditional vs. new economies), and land use behaviors (forest and agriculture vs. landscape fragmentation, residential development, and/or sprawl) that collectively can lead to rapid biodiversity losses. Positive feedbacks of information from early migrants may act to generate further in-migration, which helps drive the complex systems dynamics. Temporal lever points may exist where environmental policies (e.g., insertion of land use regulations) can act to steer a system along a new trajectory so that historical contingency and path dependence are important factors to consider. In addition to an evolutionary perspective, complexity thinking gives serious attention to the importance of space, place, and scale and heterogeneity across these domains. Some complexity proponents question the ability or goal of predictive modeling and instead propose the goal of understanding system behaviors and properties under differing scenarios, frequently using simulation modeling techniques.

Pickett et al. (2005) introduced a general biocomplexity framework for ecological investigations of coupled natural–human systems. It includes three main dimensions of complexity: (1) spatial structure, (2) organizational connectivity, and (3) temporal contingency. Themes from these dimensions are present in many of the volume chapters. Most strongly present are the dimensions of spatial structure and temporal contingency. The use of GIS and the collection and analysis of historical datasets certainly help to enable this presence. More difficult and less present is the dimension of organizational connectivity – the explicit linking and representation of how system elements interact. Observing, measuring, and representing such interactions either in quantitative models or in descriptive analytical accounts should be viewed as a major challenge facing biodiversity research encompassing coupled natural–human systems. Research funded by the US National Science Foundation's Biocomplexity in the Environment Program, initiated in 1999, reveals that the science community views integrative systems science as a promising investment. If this investment continues to yield returns, we can expect significant new understandings and policy prescriptions regarding the grand challenge of biodiversity in future years.

References

Anselin L, Syabri I, Kho Y (2006) GeoDa: an introduction to spatial data analysis. Geogr Anal 38:5–22

Blaikie PM, Brookfield HC (1987) Land degradation and society. Metheun, London

Colwell R (1998) Balancing the biocomplexity of the planet's living systems: a twenty-first century task for science. BioScience 48(10):786–787

Covich A (2000) Biocomplexity and the future: the need to unite disciplines. BioScience 50(12):1035

Crawford TW (2002) Spatial modeling of village functional territories to support population-environment linkages. In: Walsh SJ, Crews-Meyer K (eds) Linking people, place, and policy: a GIScience approach. Kluwer Academic, Boston, pp 91–111

Durraiappah AK (1998) Poverty and environmental degradation: a review and analysis of the nexus. World Dev 26(12):2169–2179

Dybas CL (2001) From biodiversity to biocomplexity: a multidisciplinary step toward understanding our environment. BioScience 51(6):426–430

Entwisle B, Stern PC (eds) (2005) Population, land use, and environment: research directions. The National Academies Press, Washington, DC

Evans TP, Moran EF (2002) Spatial integration of social and biophysical factors related to land cover change. Popul Dev Rev 28:165–186

Fotheringham AS, Brundson C, Charton M (2002) Geographically weighted regression: the analysis of spatially varying relationships. Wiley, New York

Fox J, Rindfuss RR, Walsh SJ, Mishra V (eds) (2003) People and the environment: approaches for linking household and community surveys to remote sensing and GIS. Kluwer Academic, Boston

Geoghehan J (1998) "Socializing the pixel" and "pixelizing the social" in land-use and land-cover change. In: Liverman D, Moran EF, Rindfuss RR, Stern PC (eds) People and pixels: linking remote sensing and social science. National Academy Press, Washington, DC

Gorenflo LJ, Brandon K (2006) Key human dimensions of gaps in global biodiversity conservation. BioScience 56:723–731

Harte J (2007) Human population as a dynamic factor in environmental degradation. Popul Environ 28:223–236

Heilig GK (1994) Neglected dimensions of global land-use change: reflections and data. Popul Dev Rev 20(4):831–859

Lambin EF, Baulies X, Bockstael N, Fischer G, Krug T, Leemans R, Moran EF, Rindfuss RR, Sato Y, Skole D, Turner II BL, Vogel C (1999) Land-use and land-cover change (LUCC) implementation strategy. IGBP Report 48, Royal Swedish Academy of Sciences, Stockholm, Sweden

Liu J, Dietz T, Carpenter SR, Folke C, Alberti M, Redman CL, Schneider SH, Ostrom E, Pell AN, Lubchenco J, Taylor WW, Ouyang Z, Deadman P, Kratz T, Provencher W (2007) Coupled human and natural systems. Ambio 36(8):639–649

Liverman D, Moran EF, Rindfuss RR, Stern PC (eds) (1998) People and pixels: linking remote sensing and social science. National Academy Press, Washington, DC

Marandola E, Hogan DJ (2006) Vulnerabilities and risks in population and environment studies. Popul Environ 28:83–112

Michener WK, Baerwald TJ, Firth P, Palmer MA, Rosenberger JL, Sandlin EA, Zimmerman H (2001) Defining and unraveling biocomplexity. BioScience 51(12):1018–1023

NRC (2001) Grand challenges in environmental sciences. National Academy Press, Washington, DC

Pickett STA, Cadenasso ML, Grove JM (2005) Biocomplexity in couple natural-human systems: a multidimensional framework. Ecosystems 8:225–232

Plane DA, Rogerson PA (1994) The geographical analysis of population. Wiley, New York

Rain DR, Long JF, Ratcliffe MR (2007) Measuring population pressure on the landscape: comparative GIS Studies in China, India, and the United States. Popul Environ 28:321–336

Rey SJ, Janikas MV (2006) STARS: space-time analysis of regional systems. Geogr Anal 38: 67–86

Rindfuss RR, Entwisle B, Walsh SJ, Pasartkul P, Sawangdee Y, Crawford TW, Reade J (2002) Continuous and discrete: where they have met in Nang Rong, Thailand. In: Walsh SJ, Crews-Meyer K (eds) Linking people, place, and policy: a GIScience approach. Kluwer Academic, Boston, pp 7–37

Rindfuss RR, Walsh SJ, Turner BL II, Fox J, Mishra V (2004) Developing a science of land use change: challenges and methodological issues. Proc Natl Acad Sci USA 101(39):13976–13981

Robinson WS (1950) Ecological correlations and the behavior of individuals. Am Sociol Rev 15: 351–357

Walsh SJ, Crews-Meyer KA (eds) (2002) Linking people, place, and policy: a GIScience approach. Kluwer Academic, Boston, MA

Wilson EO (2002) The future of life. Knopf, New York

Wilson EO (2005) Reflections on the future of life. In: Brown AB, Poremski K (eds) Roads to reconciliation: conflict and dialogue in the twenty-first century. ME Sharpe, Boston, pp 135–146

Index

R.P. Cincotta and L.J. Gorenflo (eds.), *Human Population: Its Influences on Biological Diversity*, Ecological Studies 214,
DOI 10.1007/978-3-642-16707-2,